T0269085

LONDON MATHEMATICAL SOCIETY LECTURE NOTE SERIES

Managing Editor: Professor J.W.S. Cassels, Department of Pure Mathematics and Mathematical Statistics, University of Cambridge, 16 Mill Lane, Cambridge CB2 1SB, England

The books in the series listed below are available from booksellers, or, in case of difficulty, from Cambridge University Press.

34 Representation theory of Lie groups, M.F. ATIYAH *et al*
36 Homological group theory, C.T.C. WALL (ed)
39 Affine sets and affine groups, D.G. NORTHCOTT
46 p-adic analysis: a short course on recent work, N. KOBLITZ
49 Finite geometries and designs, P. CAMERON, J.W.P. HIRSCHFELD & D.R. HUGHES (eds)
50 Commutator calculus and groups of homotopy classes, H.J. BAUES
57 Techniques of geometric topology, R.A. FENN
59 Applicable differential geometry, M. CRAMPIN & F.A.E. PIRANI
66 Several complex variables and complex manifolds II, M.J. FIELD
69 Representation theory, I.M. GELFAND *et al*
74 Symmetric designs: an algebraic approach, E.S. LANDER
76 Spectral theory of linear differential operators and comparison algebras, H.O. CORDES
77 Isolated singular points on complete intersections, E.J.N. LOOIJENGA
79 Probability, statistics and analysis, J.F.C. KINGMAN & G.E.H. REUTER (eds)
80 Introduction to the representation theory of compact and locally compact groups, A. ROBERT
81 Skew fields, P.K. DRAXL
82 Surveys in combinatorics, E.K. LLOYD (ed)
83 Homogeneous structures on Riemannian manifolds, F. TRICERRI & L. VANHECKE
86 Topological topics, I.M. JAMES (ed)
87 Surveys in set theory, A.R.D. MATHIAS (ed)
88 FPF ring theory, C. FAITH & S. PAGE
89 An F-space sampler, N.J. KALTON, N.T. PECK & J.W. ROBERTS
90 Polytopes and symmetry, S.A. ROBERTSON
91 Classgroups of group rings, M.J. TAYLOR
92 Representation of rings over skew fields, A.H. SCHOFIELD
93 Aspects of topology, I.M. JAMES & E.H. KRONHEIMER (eds)
94 Representations of general linear groups, G.D. JAMES
95 Low-dimensional topology 1982, R.A. FENN (ed)
96 Diophantine equations over function fields, R.C. MASON
97 Varieties of constructive mathematics, D.S. BRIDGES & F. RICHMAN
98 Localization in Noetherian rings, A.V. JATEGAONKAR
99 Methods of differential geometry in algebraic topology, M. KAROUBI & C. LERUSTE
100 Stopping time techniques for analysts and probabilists, L. EGGHE
101 Groups and geometry, ROGER C. LYNDON
103 Surveys in combinatorics 1985, I. ANDERSON (ed)
104 Elliptic structures on 3-manifolds, C.B. THOMAS
105 A local spectral theory for closed operators, I. ERDELYI & WANG SHENGWANG
106 Syzygies, E.G. EVANS & P. GRIFFITH
107 Compactification of Siegel moduli schemes, C-L. CHAI
108 Some topics in graph theory, H.P. YAP
109 Diophantine analysis, J. LOXTON & A. VAN DER POORTEN (eds)
110 An introduction to surreal numbers, H. GONSHOR
111 Analytical and geometric aspects of hyperbolic space, D.B.A. EPSTEIN (ed)
113 Lectures on the asymptotic theory of ideals, D. REES
114 Lectures on Bochner-Riesz means, K.M. DAVIS & Y-C. CHANG
115 An introduction to independence for analysts, H.G. DALES & W.H. WOODIN
116 Representations of algebras, P.J. WEBB (ed)
117 Homotopy theory, E. REES & J.D.S. JONES (eds)
118 Skew linear groups, M. SHIRVANI & B. WEHRFRITZ
119 Triangulated categories in the representation theory of finite-dimensional algebras, D. HAPPEL
121 Proceedings of *Groups - St Andrews 1985*, E. ROBERTSON & C. CAMPBELL (eds)
122 Non-classical continuum mechanics, R.J. KNOPS & A.A. LACEY (eds)
124 Lie groupoids and Lie algebroids in differential geometry, K. MACKENZIE
125 Commutator theory for congruence modular varieties, R. FREESE & R. MCKENZIE
126 Van der Corput's method of exponential sums, S.W. GRAHAM & G. KOLESNIK
127 New directions in dynamical systems, T.J. BEDFORD & J.W. SWIFT (eds)
128 Descriptive set theory and the structure of sets of uniqueness, A.S. KECHRIS & A. LOUVEAU
129 The subgroup structure of the finite classical groups, P.B. KLEIDMAN & M.W.LIEBECK
130 Model theory and modules, M. PREST

131 Algebraic, extremal & metric combinatorics, M-M. DEZA, P. FRANKL & I.G. ROSENBERG (eds)
132 Whitehead groups of finite groups, ROBERT OLIVER
133 Linear algebraic monoids, MOHAN S. PUTCHA
134 Number theory and dynamical systems, M. DODSON & J. VICKERS (eds)
135 Operator algebras and applications, 1, D. EVANS & M. TAKESAKI (eds)
136 Operator algebras and applications, 2, D. EVANS & M. TAKESAKI (eds)
137 Analysis at Urbana, I, E. BERKSON, T. PECK, & J. UHL (eds)
138 Analysis at Urbana, II, E. BERKSON, T. PECK, & J. UHL (eds)
139 Advances in homotopy theory, S. SALAMON, B. STEER & W. SUTHERLAND (eds)
140 Geometric aspects of Banach spaces, E.M. PEINADOR and A. RODES (eds)
141 Surveys in combinatorics 1989, J. SIEMONS (ed)
142 The geometry of jet bundles, D.J. SAUNDERS
143 The ergodic theory of discrete groups, PETER J. NICHOLLS
144 Introduction to uniform spaces, I.M. JAMES
145 Homological questions in local algebra, JAN R. STROOKER
146 Cohen-Macaulay modules over Cohen-Macaulay rings, Y. YOSHINO
147 Continuous and discrete modules, S.H. MOHAMED & B.J. MÜLLER
148 Helices and vector bundles, A.N. RUDAKOV *et al*
149 Solitons nonlinear evolution equations & inverse scattering, M. ABLOWITZ & P. CLARKSON
150 Geometry of low-dimensional manifolds 1, S. DONALDSON & C.B. THOMAS (eds)
151 Geometry of low-dimensional manifolds 2, S. DONALDSON & C.B. THOMAS (eds)
152 Oligomorphic permutation groups, P. CAMERON
153 L-functions and arithmetic, J. COATES & M.J. TAYLOR (eds)
154 Number theory and cryptography, J. LOXTON (ed)
155 Classification theories of polarized varieties, TAKAO FUJITA
156 Twistors in mathematics and physics, T.N. BAILEY & R.J. BASTON (eds)
157 Analytic pro-*p* groups, J.D. DIXON, M.P.F. DU SAUTOY, A. MANN & D. SEGAL
158 Geometry of Banach spaces, P.F.X. MÜLLER & W. SCHACHERMAYER (eds)
159 Groups St Andrews 1989 volume 1, C.M. CAMPBELL & E.F. ROBERTSON (eds)
160 Groups St Andrews 1989 volume 2, C.M. CAMPBELL & E.F. ROBERTSON (eds)
161 Lectures on block theory, BURKHARD KÜLSHAMMER
162 Harmonic analysis and representation theory for groups acting on homogeneous trees, A. FIGA-TALAMANCA & C. NEBBIA
163 Topics in varieties of group representations, S.M. VOVSI
164 Quasi-symmetric designs, M.S. SHRIKANDE & S.S. SANE
165 Groups, combinatorics & geometry, M.W. LIEBECK & J. SAXL (eds)
166 Surveys in combinatorics, 1991, A.D. KEEDWELL (ed)
167 Stochastic analysis, M.T. BARLOW & N.H. BINGHAM (eds)
168 Representations of algebras, H. TACHIKAWA & S. BRENNER (eds)
169 Boolean function complexity, M.S. PATERSON (ed)
170 Manifolds with singularities and the Adams-Novikov spectral sequence, B. BOTVINNIK
171 Squares, A.R. RAJWADE
172 Algebraic varieties, GEORGE R. KEMPF
173 Discrete groups and geometry, W.J. HARVEY & C. MACLACHLAN (eds)
174 Lectures on mechanics, J.E. MARSDEN
175 Adams memorial symposium on algebraic topology 1, N. RAY & G. WALKER (eds)
176 Adams memorial symposium on algebraic topology 2, N. RAY & G. WALKER (eds)
177 Applications of categories in computer science, M.P. FOURMAN, P.T. JOHNSTONE, & A.M. PITTS (eds)
178 Lower K- and L-theory, A. RANICKI
179 Complex projective geometry, G. ELLINGSRUD, C. PESKINE, G. SACCHIERO & S.A. STRØMME (eds)
180 Lectures on ergodic theory and Pesin theory on compact manifolds, M. POLLICOTT
181 Geometric group theory I, G.A. NIBLO & M.A. ROLLER (eds)
182 Geometric group theory II, G.A. NIBLO & M.A. ROLLER (eds)
183 Shintani zeta functions, A. YUKIE
184 Arithmetical functions, W. SCHWARZ & J. SPILKER
185 Representations of solvable groups, O. MANZ & T.R. WOLF
186 Complexity: knots, colourings and counting, D.J.A. WELSH
187 Surveys in combinatorics, 1993, K. WALKER (ed)
190 Polynomial invariants of finite groups, D.J. BENSON
191 Finite geometry and combinatorics, F. DE CLERCK *et al*
192 Symplectic geometry, D. SALAMON (ed)
197 Two-dimensional homotopy and combinatorial group theory, C. HOG-ANGELONI W. METZLER & A.J. SIERADSKI (eds)
198 The algebraic characterization of geometric 4-manifolds, J.A. HILLMAN

London Mathematical Society Lecture Note Series. 198

The Algebraic Characterization of Geometric 4-Manifolds

J. A. Hillman
University of Sydney

CAMBRIDGE UNIVERSITY PRESS
Cambridge, New York, Melbourne, Madrid, Cape Town, Singapore, São Paulo

Cambridge University Press
The Edinburgh Building, Cambridge CB2 2RU, UK

Published in the United States of America by Cambridge University Press, New York

www.cambridge.org
Information on this title: www.cambridge.org/9780521467780

© Cambridge University Press 1994

This publication is in copyright. Subject to statutory exception
and to the provisions of relevant collective licensing agreements,
no reproduction of any part may take place without
the written permission of Cambridge University Press.

First published 1994

A catalogue record for this publication is available from the British Library

ISBN-13 978-0-521-46778-0 paperback
ISBN-10 0-521-46778-0 paperback

Transferred to digital printing 2005

Contents

Preface

It is well known that every closed surface admits a geometry of constant curvature, and that such surfaces may be classified up to homeomorphism either by their fundamental group or by their Euler characteristic and orientation character. Much current research in dimension 3 is guided by the expectation that all closed 3-manifolds have decompositions into geometric pieces, and that (lens spaces aside) the homeomorphism type is essentially determined by the fundamental group. (Here the Euler characteristic is always 0).

In dimension 4 there is no reason to expect that every closed 4-manifold may have a geometric decomposition, and the Euler characteristic and fundamental group are independent invariants. Nevertheless the closed 4-manifolds which admit geometries or fibre over a geometric base with geometric fibre form a large and interesting class. In these notes we shall attempt to characterize algebraically such 4-manifolds (up to homotopy equivalence or homeomorphism). This task has three main parts: finding complete invariants for the homotopy type, determining which systems of invariants are realizable and applying surgery (where possible) to obtain s-cobordisms. In many cases the Euler characteristic, fundamental group and Stiefel-Whitney classes together form a complete system of invariants for the manifold and the possible invariants can be described explicitly. We shall also see that such bundle spaces usually have the minimal Euler characteristic for their fundamental groups. (The only exceptions have base and fibre S^2 or RP^2).

Our results are most satisfactory for infrasolvmanifolds. We show that a closed 4-manifold M is homeomorphic to an infrasolvmanifold if and only if $\chi(M) = 0$ and $\pi_1(M)$ has a locally nilpotent normal subgroup of Hirsch length at least 3, and two such manifolds are homeomorphic if and only if their fundamental groups are isomorphic. Moreover $\pi_1(M)$ is then a torsion free virtually poly-Z group of Hirsch length 4 and every such group is the fundamental group of an infrasolvmanifold. We also consider in detail the question of when such a manifold is the mapping torus of a self homeomorphism of a 3-dimensional infrasolvmanifold.

Typeset by $\mathcal{A}\mathcal{M}\mathcal{S}$-TEX

We also show that a closed 4-manifold M is simple homotopy equivalent to the total space of a F-bundle over B (where B and F are closed surfaces, B is aspherical and F is aspherical or S^2) if and only if $\chi(M) = \chi(B)\chi(F)$ and $\pi_1(M)$ is an extension of $\pi_1(B)$ by a normal subgroup isomorphic to $\pi_1(F)$. Any such extension is the fundamental group of such a bundle space; the bundle is determined by the group in the aspherical cases and by the group and Stiefel-Whitney classes if the fibre is S^2. The total spaces of orientable S^2-bundles over aspherical orientable surfaces are determined up to s-cobordism by theses invariants; in general there are only finitely many s-cobordism classes of such manifolds homotopy equivalent to a given bundle space.

For aspherical geometric 4-manifolds other than infrasolvmanifolds there is as yet no good intrinsic characterization of the groups arising as fundamental groups. Moreover in some other cases (notably the nonorientable total spaces of surface bundles over RP^2) we do not yet have a complete system of invariants for the homotopy type. In a recent book H.J.Baues has developed the notion of "quadratic crossed complex" and used it to show how in principle all 4-dimensional homotopy types may be classified. It should be possible to apply his ideas in these cases, but computations using quadratic crossed complexes have only been carried through for a handful of examples so far.

The major difficulty in extending this work to a classification up to homeomorphism of all such 4-manifolds is that we do not know whether s-cobordisms between 4-manifolds are always topologically products. This is only known to be so when the fundamental group is elementary amenable. Even under the latter assumption we do not know the Whitehead groups or surgery obstruction groups in most cases when the fundamental group has torsion.

The organization of this book is as follows. The first chapter is purely algebraic; here we develop the notions of elementary amenable group and safe extension of a group ring and criteria for the vanishing of cohomology of a group with coefficients in a free module. The next three chapters are homotopy theoretic. Chapter II gives general criteria for two closed 4-manifolds to be homotopy equivalent, Chapter III considers criteria for a closed 4-manifold to be homotopy equivalent to the total space of a bundle with base or fibre a circle and Chapter IV considers surface bundles. Whitehead groups and the surgery exact sequence are discussed in Chapter V. Chapters VI-IX consider the different 4-dimensional geometries, grouped according to whether the model is homeomorphic to R^4, $S^2 \times R^2$, $S^3 \times R$ or is compact. In the final chapter these results are applied to determine when the 4-manifold obtained by surgery on a 2-knot admits a geometry or a complex analytic structure, or is simple homotopy equivalent to such a manifold, and to give characterizations of minimal properly elliptic surfaces and ruled surfaces (fibred or ruled over curves of genus > 1). There is an appendix summarizing relevant properties of PD_3-complexes and a short list of open questions before the references.

I would like to thank Tom Farrell for his advice on Lemmas IV.2 and V.2, Ian Hambleton for his advice on Theorem IV.13 and Warren Dicks and Mike Mihalik for their advice on ends. I would also like to thank Michael Farber, Cherry Kearton and Jerry Levine and the Departments of Mathematics at Tel-Aviv University, the University of Durham and Brandeis University (respectively) for their hospitality and support while parts of the work presented here were done. (The balance was done at Macquarie University and the University of Sydney under the traditional conditions of academic life - that is, without external funding).

The text was prepared using $\mathcal{A}\mathcal{M}\mathcal{S}$-TEX.

Jonathan Hillman

School of Mathematics and Statistics
The University of Sydney

CHAPTER I

ALGEBRAIC PRELIMINARIES

The key algebraic idea used in this book is to study the homology of covering spaces as modules over the group ring of the group of covering transformations. In this chapter we shall summarize the relevant notions from group theory: elementary amenable groups, finiteness conditions, the stable invariant basis number property, and the connection between ends and the vanishing of cohomology with coefficients in a free module.

Our principal references for group theory are [Bi], [DD] and [Ro].

1. Group theoretic notation and terminology

Let G be a group. Then G' and ζG denote the commutator subgroup and centre of G, respectively. The *outer automorphism group* of G is $Out(G) = Aut(G)/Inn(G)$, where $Inn(G) \cong G/\zeta G$ is the subgroup of $Aut(G)$ consiting of conjugations by elements of G. If H is a subgroup of G let $N_G(H)$ and $C_G(H)$ denote the normalizer and centralizer of H in G, respectively.

If $p : G \to Q$ is an epimorphism with kernel N we shall say that G is an *extension of $Q = G/N$ by the normal subgroup N*. The action of G on N by conjugation determines a homomorphism from G/N to $Out(N) = Aut(N)/Inn(N)$. If $G/N \cong Z$ the extension splits: a choice of element t in G which projects to a generator of G/N determines a right inverse to p. Let θ be the automorphism of N determined by conjugation by t in G. Then G is isomorphic to the semidirect product $N \times_\theta Z$. Every automorphism of N arises in this way, and automorphisms whose images in $Out(N)$ are conjugate determine isomorphic semidirect products. In particular, if θ is an inner automorphism then $G \cong N \times Z$.

If P and Q are classes of groups let PQ denote the class of ("P by Q") groups G which have a normal subgroup H in P such that the quotient G/H

Typeset by $\mathcal{A}_\mathcal{M}\mathcal{S}$-TEX

is in Q, and let ℓP denote the class of ("*locally-P*") groups such that each finitely generated subgroup is contained in some P-subgroup. In particular, if F is the class of finite groups ℓF is the class of *locally finite* groups. Let *poly-P* be the class of groups with a finite composition series such that each subquotient is in P. Thus if Ab is the class of abelian groups poly-Ab is the class of solvable groups.

A group ***virtually*** has some property if it has a normal subgroup of finite index with that property. Let vP be the class of groups which are virtually P. Thus a ***virtually poly-Z*** group is one which has a subgroup of finite index with a composition series whose factors are all infinite cyclic. The number of infinite cyclic factors is independent of the choice of finite index subgroup or composition series, and is called the *Hirsch length* of the group. We shall later extend the group theoretic terminology to covering spaces.

The *Hirsch-Plotkin radical* $\sqrt{G}$ is the maximal locally-nilpotent normal subgroup of G; in a virtually poly-Z group every subgroup is finitely generated, and so $\sqrt{G}$ is then the maximal nilpotent normal subgroup. If H is normal in G then $\sqrt{H}$ is normal in G also, since it is a characteristic subgroup of H, and in particular it is a subgroup of $\sqrt{G}$.

2. Elementary amenable groups and Hirsch length

The class of *elementary amenable* groups is the class of groups generated from the class of finite groups and Z by the operations of extension and increasing union. This class arose first in connection with the Banach-Tarski paradox, but is of interest here as the largest class of groups over which topological surgery techniques are known to work in dimension 4. (We shall occasionally refer to the more general notion of amenable group. See [P]).

We may construct this class as follows. Let $X_0 = 1$ and $X_1 = AbF$ be the class of finitely generated virtually abelian groups. If X_α has been defined for some ordinal α let $X_{\alpha+1} = (\ell X_\alpha)X_1$ and if X_α has been defined for all ordinals less than some limit ordinal β let $X_\beta = \cup_{\alpha<\beta}X_\alpha$. Then the class of elementary amenable groups is $EA = \cup X_\alpha$, where the union is taken over all ordinals α.

The class EA is well adapted to arguments by transfinite induction on the ordinal $\alpha(G) = \min\{\alpha | G \in X_\alpha\}$. It is closed under extension (in fact $X_\alpha X_\beta \subseteq X_{\alpha+\beta}$) and increasing union, and under the formation of sub- and quotient groups. Moreover, if κ is the first uncountable ordinal then every countable elementary amenable group is in X_κ. Hence $EA = X_{\kappa+1} = \ell X_\kappa$. Torsion groups in EA are locally finite and no group in EA has a nonabelian free subgroup. Clearly every locally-finite by virtually solvable group is elementary amenable, i.e., $(\ell F)vpoly\text{-}Ab \subset EA$. However there are finitely generated torsion free elementary amenable groups which are not virtually solvable.

The notion of Hirsch length (as a measure of the size of a solvable group) may be extended to elementary amenable groups. The *Hirsch length* $h(G)$ of such a group G is a nonnegative integer or ∞, defined as follows. If G is in X_1 let $h(G)$ be the rank of an abelian subgroup of finite index in G. If $h(G)$ has been defined for all G in X_α and H is in ℓX_α let $h(H) = \text{l.u.b.}\{h(F) | F \leq H, F \in X_\alpha\}$. Finally, if G is in $X_{\alpha+1}$, so has a normal subgroup H in ℓX_α with G/H in X_1, let $h(G) = h(H) + h(G/H)$. Transfinite induction on $\alpha(G)$ may be used to prove (simultaneously) that h is well defined, that if H is a subgroup of G then $h(H) \leq h(G)$, that if H is a normal subgroup then $h(G) = h(H) + h(G/H)$ and that $h(G) = \text{l.u.b.}\{h(F) | F \textit{ is a finitely generated subgroup of } G\}$ [Hi91].

Lemma 1. *Let G be a finitely generated infinite elementary amenable group. Then G has normal subgroups $K < H$ such that G/H is finite, H/K is free abelian of positive rank and the action of G/H on H/K by conjugation is effective.*

Proof. We may show that G has a normal subgroup K such that G/K is an infinite virtually abelian group, by transfinite induction on $\alpha(G)$. We may assume that G/K has no nontrivial finite normal subgroup. If H is a subgroup of G which contains K and is such that H/K is a maximal abelian normal subgroup of G/K then H and K satisfy the above conditions. //

Lemma 2. *Let $G = \cup_{n \geq 0} G_n$ be a group which the union of an increasing sequence of subgroups G_n such that each subgroup G_n has a maximal solvable normal subgroup H_n of derived length at most d and index at most M. Then G has a maximal solvable normal subgroup, of derived length at most $d + 2M$*

and index at most $M!$.

Proof. Define an increasing sequence of subgroups $\bar{H}_n$ such that $H_i \leq \bar{H}_i \leq G_i$ for all $i \geq 0$ by $\bar{H}_0 = H_0$ and $\bar{H}_{j+1} = \bar{H}_j H_{j+1}$. It is easily seen by induction that for each $j \geq 0$ the subgroup $\bar{H}_j$ is a solvable subgroup of derived length at most $d+M$ and index at most M in G_j. (Note that an increasing union of solvable subgroups of derived length at most $d+M$ is solvable and of derived length at most $d+M$). Similarly, $H = \cup_{j\geq 0} H_j$ is a solvable subgroup of G which is of derived length at most $d+M$. Any finite set of coset representatives for H in G must lie in some common subgroup G_j and be in distinct cosets of H_j there. Therefore the index of H in G is at most M, and so the intersection of the conjugates of H in G is a solvable normal subgroup of index at most $M!$. Therefore G has a maximal normal solvable subgroup, S say, of index at most $M!$. Since S is an extension of $S/H\cap S$ by $H\cap S$ it has derived length at most $M+d+M = d+2M$. //

Theorem 3. *Let G be a countable torsion free elementary amenable group. If $h(G) < \infty$ then G is virtually solvable.*

Proof. We shall show by induction on h that there are functions d and M from the set of nonnegative integers to itself such that every countable torsion free elementary amenable group of Hirsch length $h < \infty$ has a maximal solvable normal subgroup of derived length at most $d(h)$ and index at most $M(h)$. Since the only such group of Hirsch length 0 is the trivial group we may set $d(0) = 0$ and $M(0) = 1$. Suppose that the result is true for all such groups with Hirsch length at most h. If G has Hirsch length $h+1$ and is finitely generated then by Lemma 1 it has normal subgroups $K < H$ such that G/H is finite, H/K is free abelian of rank $r \geq 1$ and the action of G/H on H/K by conjugation is effective. In particular, G/H is isomorphic to a finite subgroup of $GL(r, Z)$. Since the kernel of the reduction of coefficients homomorphism from $GL(r, Z)$ to $GL(r, Z/pZ)$ is torsion free for all odd primes p it follows that G/H has order at most 3^{r^2}. As K is torsion free and elementary amenable and $k = h(K) = h_1 - r \leq h$ it has a maximal solvable normal subgroup, L say, of derived length at most $d(k)$ and index at most $M(k)$, by the hypothesis of induction. Since L is characteristic in K it is normal in G. The quotient group

G/L has a free abelian normal subgroup of index at most $[G:H]+[K:L]!$. Let $M' = \max\{3^{r^2}+M(k)!|0 \leq k \leq h, r = h+1-k\}$ and $d' = d(h)+1+M'!$. Then the maximal solvable normal subgroup of G has derived length at most d' and index at most M'. As any countable group is the union of an increasing sequence of finitely generated subgroups the general case follows from Lemma 2 on setting $d(h+1) = d'+2M'$ and $M(h+1) = M'!$. //

It can be shown that $h(G) < \infty$ if and only if G has normal subgroups $K \leq H \leq G$ such that K is locally-finite, H/K is solvable and of finite Hirsch length and G/H is finite [HL92]. A virtually solvable group of finite Hirsch length and with no nontrivial locally-finite normal subgroup must be countable, by Lemma 7.9 of [Bi].

Lemma 4. *Let G be an elementary amenable group. If $h(G) = \infty$ then for every $k > 0$ there is a subgroup H of G with $k < h(H) < \infty$.*

Proof. We shall argue by induction on $\alpha(G)$. The result is vacuously true if $\alpha(G) = 1$. Suppose that it is true for all groups in X_α and G is in ℓX_α. Since $h(G) = \mathrm{l.u.b.}\{h(F)|F \leq G, F \in X_\alpha\}$ either there is a subgroup F of G in X_α with $h(F) = \infty$, in which case the result is true by the inductive hypothesis, or $h(G)$ is the least upper bound of a set of natural numbers and the result is true. If G is in $X_{\alpha+1}$ then it has a normal subgroup N which is in ℓX_α with quotient G/N in X_1. But then $h(N) = h(G) = \infty$ and so N has such a subgroup. //

Theorem 5. *Let G be a countable elementary amenable group of finite cohomological dimension. Then $h(G) \leq c.d.G$ and G is virtually solvable.*

Proof. Since $c.d.G < \infty$ the group G is torsion free. Let H be a subgroup of finite Hirsch length. Then H is virtually solvable and $c.d.H \leq c.d.G$ so $h(H) \leq c.d.G$. The theorem now follows from Theorem 3 and Lemma 4. //

The assumption that G be countable is unnecessary. (See [HL92]).

3. Modules and finiteness conditions

Let G be a group and R a commutative ring. If $w : G \to Z^\times = \{\pm 1\} \cong Z/2Z$ is a homomorphism then $\bar{g} = w(g)g^{-1}$ defines an anti-involution on

$R[G]$. If L is a left $R[G]$-module $\bar{L}$ shall denote the *conjugate* right $R[G]$-module with the same underlying R-module and $R[G]$-action given by $l.g = \bar{g}.l$, for all $l \in L$ and $g \in G$. (We shall also use the overbar to denote the conjugate of a right $R[G]$-module). The conjugate of a free left (right) module is a free right (left) module of the same rank.

We shall also let Z^w denote the G-module with underlying abelian group Z and G-action given by $g.n = w(g)n$ for all g in G and n in Z.

Lemma 6. [Wl65] *Let G and H be groups such that G is finitely presentable and there are homomorphisms $j : H \to G$ and $\rho : G \to H$ with $\rho j = id_H$. Then H is also finitely presentable.*

Proof. Since G is finitely presentable there is an epimorphism $p : F \to G$ from a free group $F(X)$ with a finite basis X onto G, with kernel the normal closure of a finite set of relators R. We may choose elements w_x in $F(X)$ such that $j\rho p(x) = p(w_x)$, for all x in X. Then ρ factors through the group K with presentation $< X|R, x^{-1}w_x, \forall x \in X >$, say $\rho = vu$. Now uj is clearly onto, while $vuj = \rho j = id_H$, and so v and uj are mutually inverse isomorphisms. Therefore $H \cong K$ is finitely presentable. //

A group G is FP_n if the augmentation $Z[G]$-module Z has a projective resolution which is finitely generated in degrees $\leq n$. It is FP if it has finite cohomological dimension and is FP_n for $n = c.d.G$; it is FF if moreover Z has a finite resolution consisting of finitely generated free $Z[G]$-modules. "Finitely generated" is equivalent to FP_1, while "finitely presentable" implies FP_2. Groups which are FP_2 are also said to be *almost finitely presentable.* (It remains unkown whether FP_2 groups are finitely presentable).

If the augmentation $Q[\pi]$-module Q has a finite resolution F_* by finitely generated free modules then the alternating sum $\chi(\pi) = \Sigma(-1)^i rank(F_i)$ is independent of the resolution. (If π is the fundamental group of an aspherical finite complex K then $\chi(\pi) = \chi(K)$). This definition may be extended to groups σ which have a subgroup π of finite index with such a resolution by setting $\chi(\sigma) = \chi(\pi)/[\sigma : \pi]$. (It is not hard to see that this is well defined [Se71]).

4. The SIBN property and safe extensions of group rings

Kropholler, Linnell and Moody have shown that if H is an elementary amenable group whose finite subgroups have bounded order and which has no nontrivial finite normal subgroup then the group ring $Z[H]$ has a classical ring of fractions which is a matrix ring over a division ring [KLM88]. That is, there is a division ring D and an embedding $i : Z[H] \to M_n(D)$ (where n is the least common multiple of the orders of the finite subgroups of H) such that the image of every nonzero divisor of $Z[G]$ is invertible in $M_n(D)$ and every element of $M_n(D)$ can be uniquely expressed in the form $i(\delta)^{-1}i(\gamma)$ for some γ and δ in $Z[H]$. Since every finitely generated free left $M_n(D)$-module is a finite dimensional left D-vector space, every onto endomorphism of such a module is an isomorphism. A ring R for whch this holds is said to have the *strong invariant basis number* (SIBN) property. Equivalently, a ring R has the SIBN property if whenever an R-module L satisfies $L \oplus R^a \cong R^b$ for some nonnegative integers a, b then $b - a$ depends only on L and is 0 if and only if $L = 0$. (Note that if a ring has the SIBN property then so does any subring).

Kaplansky showed that group rings have the SIBN property (see page 122 of [K]). Rosset extended his argument, to show that if a group G has a torsion free abelian normal subgroup A then the multiplicative system $S = C[A]\backslash\{0\}$ is an Ore system in $C[G]$, and the localization $C[G]_S$ is a safe extension of $Z[G]$ [Ro84]. We shall need a further extension of this result, due essentially to Linnell. If G is a group and $w : G \to Z/2Z$ is a homomorphism we shall say that an extension of rings $Z[G] \subseteq \Phi$ is a *safe extension* if Φ has an involution extending that of $Z[G]$, Φ has the SIBN property, Φ is flat as a right $Z[G]$=module and $\Phi \otimes_{Z[G]} Z = 0$.

Theorem 7. [Li91] *Let G be a group. Then $Z[G]$ has the SIBN property. If G has a nontrivial elementary amenable normal subgroup N whose finite subgroups have bounded order and which has no nontrivial finite normal subgroup then $Z[G]$ has a safe extension.*

Proof. We shall only outline the argument of Kaplansky, Rosset and Linnell, refering to [Ro84] and [Li91] for further details.

Let $B = C[G]$ and equip B with the involution given by $(cg)^* = \bar{c}g^{-1}$ for

c in C and g in G, and the inner product given by $(\Sigma a_g g, \Sigma b_g g) = \Sigma a_g \bar{b}_g$. Let H be the Hilbert space completion of B. Elements of $C[N]$ act by left multiplication as bounded operators on H. The function sending ξ in $C[N]$ to the corresponding operator T_ξ embeds $C[N]$ in $B(H)$, the ring of bounded operators on H. According to [Ro68] the weak closure W of $C[N]$ in $B(H)$ can be embedded in a ring $\tilde{W}$ of densely defined operator on H which has the SIBN property and is in which every principal (left) ideal is generated by a projection which lies in W. By Theorem 4 of [Li91], if ξ is a nonzerodivisor of $C[N]$ then T_ξ is injective. Now $\tilde{W}T_\xi = \tilde{W}e$ for some idempotent e in W. Since $T_\xi(1-e) = 0$ we must have $e = 1$ and so T_ξ has a left inverse. Since $\tilde{W}$ has the SIBN property it follows that T_ξ is invertible. Thus if $F = M_n(D)$ is the classical ring of fractions for $Z[N]$ it embeds in $\tilde{W}$ and so also has the SIBN property.

Since F is a direct limit of free right $Z[N]$-modules it is flat as a $Z[N]$-algebra. Therefore $F_G = F \otimes_{Z[N]} Q[G]$ is flat as a right $Z[G]$-module. We may define a multiplication which makes this module into a $Z[G]$-algebra by

$$(s_1^{-1}r_1 \otimes \alpha)(s_2^{-1}r_2 \otimes \beta) = s_1^{-1}r_1(\alpha s_2 \alpha^{-1})^{-1}(\alpha r_2 \alpha^{-1}) \otimes \alpha\beta$$

for r_1, r_2 in $Z[N]$, s_1, s_2 in $Z[N]\backslash\{0\}$ and α, β in G. We may also define an involution on F_G by $\overline{\gamma^{-1}} = (\bar{\gamma})^{-1}$.

The extended augmentation module $F_G \otimes_{Z[G]} Z = F_G \otimes_{Z[G]} Q$ is a Q-vector space, of dimension at most 1. However it is also a left D-vector space. As $Q[N]$ embeds in $F = M_n(D)$ and as N is infinite D has infinite dimension over Q. Therefore $F_G \otimes_{Z[G]} Z$ must be 0. That F_G has the SIBN property follows as in [Ro84] on using Theorem 4 of [Li91] instead of Rosset's 3.4. //

If N is torsion free we may argue that if $n \neq 1$ in N then $n-1$ is a nonzero divisor in $Z[N]$ and so is invertible in F and hence in F_G. As it annihilates the augmentation module Z it follows that $F_G \otimes_{Z[G]} Z = 0$.

On the other hand, if $Z[\pi]$ has a safe extension Φ and the augmentation $Q[\pi]$-module Q has a finite free resolution F_* then on tensoring F_* with Φ we get an exact sequence of free Φ-modules, and hence $\chi(\pi) = 0$. In particular, the group ring of a noncyclic free group does not have a safe extension.

5. Ends and cohomology with free coefficients

A finitely generated group G has 0, 1, 2 or infinitely many ends. It has 0 ends if and only if it is finite, in which case $H^0(G;Z[G]) \cong Z$ and $H^q(G;Z[G]) = 0$ for $q > 0$. Otherwise $H^0(G;Z[G]) = 0$ and $H^1(G;Z[G])$ is a free abelian group of rank $e(G)-1$, where $e(G)$ is the number of ends of G [Sp49]. The group G has more than one end if and only if it is either a nontrivial generalised free product with amalgamation $G \cong A *_C B$ or an HNN extension $A *_C \phi$ where C is a finite group. In particular, it has two ends if and only if it is virtually Z if and only if it has a (maximal) finite normal subgroup F such that the quotient G/F is either infinite cyclic (Z) or infinite dihedral ($D = (Z/2Z) * (Z/2Z)$). (See [DD]).

Lemma 8. *Let N be a finitely generated elementary amenable group with $h(N) > 1$. Then N has one end.*

Proof. Any group with infinitely many ends has nonabelian free subgroups, and so cannot be amenable. If $h(N) > 1$ then N is infinite and not virtually Z, and so must have one end. //

Let G be a group with a normal subgroup N, and let A be a left $Z[G]$-module. The *Lyndon-Hochschild-Serre spectral sequence* (LHSSS) for G as an extension of G/N by N and with coefficients the $Z[G]$-module A has E_2 term $H^p(G/N;H^q(N;A))$, r^{th} differential of bidegree $(r, 1-r)$ and converging to $H^{p+q}(G;A)$. (See Section 10.1 of [Mc]).

The argument of the next result is from [Ro75].

Theorem 9. *If G has a normal subgroup N which is the union of an increasing sequence of FP_r subgroups N_n such that $H^s(N_n;Z[N_n]) = 0$ for $s \leq r$ then $H^s(G;Z[G]) = 0$ for $s \leq r$.*

Proof. Let $s \leq r$. Since N_n is FP_r we have $H^s(N_n;Z[G]) = H^s(N_n;Z[N_n]) \otimes Z[G/N_n] = 0$. Let f be an s-cocycle for N with coefficients $Z[G]$, and let f_n denote the restriction of f to a cocycle on N_n. Then there is an $(s-1)$-cochain g_n on N_n such that $\delta g_n = f_n$. Since $\delta(g_{n+1}|_{N_n} - g_n) = 0$ and $H^{s-1}(N_n;Z[G]) = 0$ there is an $(s-2)$-cochain h_n on N_n with $\delta h_n = g_{n+1}|_{N_n} - g_n$. Choose an extension h_n' of h_n to N_{n+1} and let $\bar{g}_{n+1} = g_{n+1} - \delta h_n'$.

Then $\bar{g}_{n+1}|_{N_n} = g_n$ and $\delta\bar{g}_{n+1} = f_{n+1}$. In this way we may extend g_0 to an $(s-1)$-cochain g on N such that $f = \delta g$ and so $H^s(N; Z[G]) = 0$. The LHSSS for G as an extension of G/N by N, with coefficients $Z[G]$, now gives $H^s(G; Z[G]) = 0$ for $s \leq r$. //

Corollary. *If G has a normal subgroup N which is the union of an increasing sequence of finitely generated, one-ended subgroups then G has one end.* //

In particular, this corollary applies if N is a countable elementary amenable group and $h(N) > 1$, by Lemma 8. We can prove a vanishing theorem for higher cohomology groups under more restrictive assumptions on the normal subgroup.

Theorem 10. *If G has a countable locally nilpotent normal subgroup N such that $h(N) > r$ then $H^s(G; Z[G]) = 0$ for $s \leq r$.*

Proof. The subgroup N is the union of an increasing sequence of finitely generated nilpotent subgroups of Hirsch length $> r$. As finitely generated nilpotent groups are virtually poly-Z, Theorem 9 applies. //

Does this theorem remain true if N is assumed only to be an elementary amenable group with $h(N) > r$?

The second cohomology of a group with free coefficients ($H^2(G; Z[G])$) shall play an important role in our investigations. In particular, we would like to know when this group is 0, and when it is infinite cyclic. Unfortunately less is known about this group than about $H^1(G; Z[G])$. In [Fa74] Farrell has shown that if G is finitely presentable and has an element of infinite order then $H^2(G; Z[G])$ is either 0 or Z or is not finitely generated. He has also shown that if G is finitely presentable, has one end and $H^2(G; Z/2Z[G]) = Z/2Z$ then every finitely generated subgroup with one end has finite index in G (Proposition 2.4 of [Fa74]). (In particular, if G is torsion free then subgroups of infinite index in G are locally free). The heart of the argument involves material on the cohomology of a space relative to a family of supports. It would be of interest to have a purely algebraic exposition of this work of Farrell.

Theorem 11. *Let G be a finitely presentable group such that $H^2(G; Z[G]) \cong Z$. Then G has one end.*

Proof. The group G is infinite since $H^2(G; Z[G]) \neq 0$. If it has more than one end then it is a generalised free product with amalgamation $G \cong A *_C B$ or an HNN extension $A *_C \phi$, where C is finite and $A \neq C \neq B$, by Theorem IV.6.10 of [DD]. The Mayer-Vietoris sequence for such a decomposition gives an isomorphism $H^2(G; Z[G]) \cong H^2(A; Z[G]) \oplus H^2(B; Z[G])$, since $H^s(C; W) = 0$ for all $s > 0$ and any free $Z[C]$-module W. Since G and C are FP_2 so are the groups A and B, by Proposition 2.13 of [Bi]. But then $H^2(A; Z[G])$ and $H^2(B; Z[G])$ are each either 0 or of infinite rank, since $[G : A] = [G : B] = \infty$. Hence G must have one end. //

CHAPTER II

GENERAL RESULTS ON THE HOMOTOPY TYPE OF 4-MANIFOLDS

The homotopy type of an n-manifold is largely determined by its $[(n+1)/2]$-skeleton and Poincaré duality. Thus in dimension 4 the primary invariants are the fundamental group, the orientation character and the second homotopy group; for the cases of interest to us the latter invariant is often determined by the fundamental group and the Euler characteristic. This chapter begins with a brief review of Poincaré duality and the Universal Coefficient spectral sequence. We then apply these to give criteria for two 4-manifolds to be homotopy equivalent. We also give criteria for a closed 4-manifold to be aspherical or more generally for its universal covering space to be homotopy equivalent to a finite complex. In the final section we obtain estimates for the minimal Euler characteristic of closed 4-manifolds with fundamental group of cohomological dimension at most 2 and determine the second homotopy groups of manifolds realizing the minimal value.

1. Equivariant (co)homology and Poincaré duality

Let X be a connected cell complex and let $\tilde{X}$ be its universal covering space. If H is a normal subgroup of $G = \pi_1(X)$ we may lift the cellular decomposition of X to an equivariant cellular decomposition of the corresponding covering space X_H. The cellular chain complex C_* of X_H with coefficients in a commutative ring R is then a complex of left $R[G/H]$-modules, with respect to the action of the covering group G/H. Moreover C_* is a complex of free modules, with bases obtained by choosing a lift of each cell of X. If X is a finite complex G is finitely presentable and these modules are finitely generated. If X is finitely dominated, i.e., is a retract of a finite complex Y, then G is a retract of $\pi_1(Y)$ and so is finitely presentable, by Lemma I.6. Moreover the

Typeset by $\mathcal{A}_{\mathcal{M}}\mathcal{S}$-TeX

chain complex C_* of the universal cover is chain homotopy equivalent over $R[G]$ to a complex of finitely generated projective modules [Wl65].

The ith *equivariant homology* module of X with coefficients $R[G/H]$ is the left module $H_i(X; R[G/H]) = H_i(C_*)$, which is clearly isomorphic to $H_i(X_H; R)$ as an R-module, with the action of the covering group determining its $R[G/H]$-module structure. The ith *equivariant cohomology* module of X with coefficients $R[G/H]$ is the right module $H^i(X; R[G/H]) = H^i(C^*)$, where $C^* = Hom_{R[G/H]}(C_*, R[G/H])$ is the associated cochain complex of right $R[G/H]$-modules. More generally, if A and B are right and left $Z[G/H]$-modules (respectively) we may define $H_j(X; A) = H_j(A \otimes_{Z[G/H]} C_*)$ and $H^{n-j}(X; B) = H^{n-j}(Hom_{Z[G/H]}(C_*, B))$. There is a Universal Coefficient Spectral Sequence (UCSS) which converges to $H^{p+q}(X; R[G/H])$, with E_2^{pq} term $Ext^q_{R[G/H]}(H_p(X; R[G/H]), R[G/H])$ and whose r^{th} differential d_r has bidegree $(1-r, r)$.

A *PD_n-complex* is a finitely dominated cell complex which satisfies Poincaré duality of formal dimension n with local coefficients. It is *finite* if it is homotopy equivalent to a finite cell complex. (It is most convenient for our purposes below to require that PD_n-complexes be finitely dominated. If a CW-complex X satisfies local duality then $\pi_1(X)$ is FP_2, and X is finitely dominated if and only if $\pi_1(X)$ is finitely presentable [Br72, Br75]. Thus if G is a PD_n-group in the sense of [Bi] then $K(G, 1)$ is finitely dominated if and only if G is finitely presentable. Ranicki uses the broader definition in his book [Rn]). All the PD_n-complexes that we consider shall be assumed to be connected.

Let P be a PD_n-complex and C_* be the cellular chain complex of $\tilde{P}$. Then the Poincaré duality isomorphism may also be described in terms of a chain homotopy equivalence from $\overline{C^*}$ to C_{n-*}, which induces isomorphisms from $\overline{H^j(C^*)}$ to $H_{n-j}(C_*)$, given by cap product with a generator $[P]$ of $H_n(P; Z^{w_1(P)}) = H_n(\bar{Z} \otimes_{Z[\pi_1(P)]} C_*)$. (Here the first Stiefel-Whitney class $w_1(P)$ is considered as a homomorphism from $\pi_1(P)$ to $Z/2Z$). From this point of view it is easy to see that Poincaré duality gives rise to (Z-linear) isomorphisms from $H_j(P; B)$ to $H^{n-j}(P; \bar{B})$, where B is any right $Z[\pi_1(P)]$-module of coefficients. (See [Wl67] or Chapter II of [W] for further details).

Throughout this book *closed manifold* shall mean compact, connected TOP manifold without boundary. Every closed manifold has the homotopy type of a finite Poincaré duality complex [KS].

2. Homotopy types.

If M is a cell complex $P_2(M)$ shall denote the second stage of the Postnikov tower for M, and $c_M = c_{P_2(M)}f_M$ the factorization of the classifying map $c_M : M \to K(\pi_1(M),1)$ through $f_M : M \to P_2(M)$ and $c_{P_2(M)} : P_2(M) \to K(\pi_1(M),1)$. A map $f : X \to K(\pi_1(M),1)$ lifts to a map from X to $P_2(M)$ if and only if $f^*k_1(M) = 0$, where $k_1(M)$ is the first k-invariant of M in $H^3(\pi_1(M);\pi_2(M))$. The *algebraic* 2-*type* of M is the triple $[\pi, \pi_2(M), k_1(M)]$. Two such triples $[\pi, \Pi, \kappa]$ and $[\pi', \Pi', \kappa']$ (corresponding to M and M', respectively) are equivalent if there are isomorphisms $\alpha : \pi \to \pi'$ and $\beta : (\Pi, S) \to (\Pi', (\pm)S')$ such that $\beta(gm) = \alpha(g)\beta(m)$ for all $g \in \pi$ and $m \in \Pi$ and $\beta_*\kappa = \alpha^*\kappa'$ in $H^3(\pi, \alpha^*\Pi')$. Such an equivalence may be realized by a homotopy equivalence of $P_2(M)$ and $P_2(M')$. (The reference [Ba] gives a detailed treatment of Postnikov factorizations of nonsimple maps and spaces).

Lemma 1. *Let R be a ring and C_* be a finite chain complex of projective R-modules. If $H_i(C_*) = 0$ for $i < q$ and $H^{q+1}(Hom_R(C_*, B)) = 0$ for any left R-module B then $H_q(C_*)$ is projective. If moreover $H_i(C_*) = 0$ for $i > q$ then $H_q(C_*) \oplus \bigoplus_{i\equiv q+1\ (2)} C_i \cong \bigoplus_{i\equiv q\ (2)} C_i$.*

Proof. We may assume without loss of generality that $q = 0$ and $C_i = 0$ for $i < 0$. We may factor $\partial_1 : C_1 \to C_0$ through $B = \mathrm{Im}\ \partial_1$ as $\partial_1 = j\beta$, where β is an epimorphism and j is the natural inclusion of the submodule B. Since $j\beta\partial_2 = \partial_1\partial_2 = 0$ and j is injective $\beta\partial_2 = 0$. Hence β is a 1-cocycle of the complex $Hom_R(C_*, B)$. Since $H^1(Hom_R(C_*, B)) = 0$ there is a homomorphism $\sigma : C_0 \to B$ such that $\beta = \sigma\partial_1 = \sigma j\beta$. Since β is an epimorphism $\sigma j = id_B$ and so B is a direct summand of C_0. This proves the first assertion.

The second assertion follows by an induction on the length of the complex. //

Most of the results of this section depend on the following lemma, in conjunction with use of the Euler characteristic to compute the rank of the surgery kernel. (This lemma is based on Lemmas 2.2 and 2.3 of [W]).

Lemma 2. *Let $f : M \to E$ be a map of PD_4-complexes which induces an isomorphism of fundamental groups, respects the orientation characters and such that $f_*[M] = \pm[E]$. Then the surgery kernel* $\mathrm{Ker}\pi_2(f)$ *is projective.*

Proof. Up to homotopy type we may assume that that M is a subcomplex of E. We may also identify $\pi_1(M)$ with $\pi = \pi_1(E)$. Let $C_*(M)$, $C_*(E)$ and D_* be the cellular chain complexes of $\tilde{M}$, $\tilde{E}$ and $(\tilde{E}, \tilde{M})$, respectively. Then the sequence $0 \to C_*(M) \to C_*(E) \to D_* \to 0$ is a short exact sequence of finitely generated projective $Z[\pi]$-chain complexes.

By the projection formula $f_*(f^*a \cap [M]) = a \cap f_*[M] = \pm a \cap [E]$ for any cohomology class $a \in H^*(E; Z[\pi])$. Since M and E satisfy Poincaré duality it follows that f induces split surjections on homology and split injections on cohomology. Hence $H_q(D_*)$ is the "surgery kernel" in degree $q-1$, and the duality isomorphisms induce isomorphisms from $H^r(Hom_{Z[\pi]}(D_*, B))$ to $H_{6-r}(D_* \otimes B)$, where B is any left $Z[\pi]$-module. Since f induces isomorphisms on homology and cohomology in degrees ≤ 1, with any coefficients, it follows that the hypotheses of Lemma 1 are satisfied for the $Z[\pi]$-chain complex D_*, with $q = 3$, and so $H_3(D_*) = \mathrm{Ker}\pi_2(f)$ is projective. //

Theorem 3. *Let E and M be finite PD_4-complexes. A map $f : M \to E$ is a homotopy equivalence if and only if it induces an isomorphism on π_1, $f^*w_1(E) = w_1(M)$, $f_*[M] = \pm[E]$ and $\chi(M) = \chi(E)$.*

Proof. The conditions are clearly necessary. Suppose that they hold. As in Lemma 2 we may assume that f is a cellular inclusion of finite cell complexes, and we may identify $\pi_1(M)$ with $\pi = \pi_1(E)$. Let D_* be the cellular chain complex of $(\tilde{E}, \tilde{M})$. Then D_* is a complex of finitely generated free $Z[\pi]$-modules, and $H_3(D_*) \cong \mathrm{Ker}\pi_2(f)$ is the only obstruction to f being a homotopy equivalence. Lemmas 1 and 2 together imply that $H_3(D_*) \oplus \bigoplus_{i\ odd} D_i \cong \bigoplus_{i\ even} D_i$. Thus $H_3(D_*)$ is a stably free $Z[\pi]$-module of rank $\chi(E, M) = \chi(M) - \chi(E) = 0$ and so it is trivial, as $Z[\pi]$ has the SIBN property, by Theorem I.7. Therefore f is a homotopy equivalence. //

Corollary. *If E is aspherical then M is homotopy equivalent to E if and only if there is an isomorphism $\theta : \pi_1(M) \to \pi_1(E)$ such that $w_1(M) = w_1(E)\theta$, $\theta_* c_{M*}[M] = \pm[E]$ and $\chi(M) = \chi(E)$.* //

With a little more effort we can relax the hypothesis that the classifying map have degree ± 1.

Lemma 4. *Let M be a PD_4-complex with fundamental group π. Then there is an exact sequence $0 \to \overline{H^2(\pi; Z[\pi])} \to \pi_2(M) \to \overline{Hom_{Z[\pi]}(\pi_2(M), Z[\pi])}$.*

Proof. If X is any finitely dominated cell complex with fundamental group π the UCSS gives an exact sequence $0 \to H^2(\pi; Z[\pi]) \to H^2(X; Z[\pi]) \to Hom_{Z[\pi]}(H_2(X; Z[\pi]), Z[\pi])$. The lemma follows immediately on applying the Hurewicz theorem and Poincaré duality. //

Theorem 5. *Let M be a finite PD_4-complex with fundamental group π. Then M is aspherical if and only if π is a (finitely presentable) PD_4-group of type FF, $c_M^* w_1(\pi) = w_1(M)$, $c_{M*}[M] \neq 0$ and $\chi(M) = \chi(\pi)$.*

Proof. The conditions are clearly necessary. Suppose that they hold. Then $K(\pi, 1)$ is a finite PD_4-complex. Without loss of generality we may assume that M and π are orientable and that c_M is a cellular inclusion of finite complexes. Since $\underline{H^2(\pi; Z[\pi])} = 0$ Lemma 4 gives a monomorphism from $H_2(\tilde{M}; Z)$ to $\overline{Hom_{Z[\pi]}(H_2(\tilde{M}; Z), Z[\pi])}$. Similarly there is a monomorphism from $H_2(\tilde{M}; Q)$ to $\overline{Hom_{Z[\pi]}(H_2(\tilde{M}; Q), Q[\pi])}$ and so the natural map from $H_2(\tilde{M}; Z)$ to $H_2(\tilde{M}; Q)$ is injective. Since c_M has nonzero degree, $H_2(\tilde{M}; Q)$ is the rational surgery kernel. As in Theorem 3 it is a stably free $Q[\pi]$-module of rank $\chi(M) - \chi(\pi) = 0$ and so is 0, by Theorem I.7. Hence M is aspherical. //

Some condition on the degree is necessary. Given any finitely presentable group G there is a closed orientable 4-manifold M with $\pi_1(M) \cong G$ and such that $c_{M*}[M] = 0$ in $H_4(G; Z)$. We may take M to be the boundary of a regular neighbourhood N of some embedding in R^5 of a finite 2-complex K with $\pi_1(K) \cong G$. As the inclusion of M into N is 2-connected and K is a deformation retract of N the classifying map c_M factors through c_K and so induces the trivial homomorphism on homology in degrees > 2.

The next theorem together with the observation following Theorem I.7 gives another criterion for asphericity.

Theorem 6. *Let M be a finite PD_4-complex with fundamental group π. Suppose that π has one end and that $Z[\pi]$ has a safe extension Φ. Then $H_2(M;\Phi) = \Phi \otimes_{Z[\pi]} \pi_2(M)$ is stably free of rank $\chi(M)$, and so $\chi(M) \geq 0$. In particular, if $\chi(M) = 0$ then $Hom_{Z[\pi]}(\pi_2(M), Z[\pi]) = 0$ and so $\pi_2(M) \cong \overline{H^2(\pi; Z[\pi])}$.*

Proof. Let C_* be the cellular chain complex of $\tilde{M}$ with coefficients Z. Since π has one end $H_i(C_*) = 0$ if $i \neq 0$ or 2, while $H_0(C_*) \cong Z$ and $H_2(C_*) \cong \pi_2(M)$. If B is any left Φ-module then $H^3(Hom_\Phi(\Phi \otimes_{Z[\pi]} C_*, B)) = H_1(\bar{B} \otimes_{Z[\pi]} C_*) = 0$ by Poincaré duality and the fact that $\Phi \otimes_{Z[\pi]} C_*$ is exact in degrees ≤ 1. Hence $H_2(\Phi \otimes_{Z[\pi]} C_*)$ is a stably free Φ-module of rank $\chi(M)$, by Lemma 1. Since Φ is a flat extension of $Z[\pi]$ this module is isomorphic to $\Phi \otimes_{Z[\pi]} H_2(C_*) = \Phi \otimes_{Z[\pi]} \pi_2(M)$.

If $\chi(M) = 0$ this module is 0, since Φ has the SIBN property. Therefore $Hom_{Z[\pi]}(\pi_2(M), \Phi) = Hom_\Phi(\Phi \otimes_{Z[\pi]} \pi_2(M), \Phi) = 0$. Since $Z[\pi]$ embeds in Φ this implies that $Hom_{Z[\pi]}(\pi_2(M), Z[\pi]) = 0$, and the final assertion now follows from Lemma 4. //

Corollary. *Suppose that $Z[\pi]$ has a safe extension. Then M is aspherical if and only if $\chi(M) = 0$ and $H^s(\pi; Z[\pi]) = 0$ for $s \leq 2$. If so π is a PD_4-group.* //

In one of our characterizations of infrasolvmanifolds in Chapter VI we shall need the following extension of this corollary. The argument is very similar to that of Theorem 6, and we shall not give it here. (See Theorem III.3 of [H]).

Addendum. *Let M be a finite PD_4-complex with fundamental group π. Suppose that π has a normal subgroup ν such that $Z[\pi/\nu]$ has a safe extension and that R is a subring of Q. Then the covering space M_ν with fundamental group ν is R-acyclic if and only if $\chi(M) = 0$, $H^s(\pi/\nu; R[\pi/\nu]) = 0$ for $s \leq 2$ and $H_1(\nu; R) = 0$. If so π/ν is a PD_4-group over R.* //

If π has an elementary amenable normal subgroup N of Hirsch length greater than 1 then π has one end, i.e., $H^s(\pi; Z[\pi]) = 0$ for $s \leq 1$. Does $h(N) > 2$ imply that $H^2(\pi; Z[\pi]) = 0$ also? Eckmann has shown that a closed 4-manifold M whose fundamental group π is amenable is aspherical if and only if $\chi(M) = 0$ and $H^s(\pi; Z[\pi]) = 0$ for $s \leq 2$ [Ec93]. Can the assumption that π be amenable be weakened to π has an infinite amenable normal subgroup? (In one direction the answer is yes; if the fundamental group of an aspherical finite complex has such a subgroup then the Euler characteristic is 0 [CG86]).

The hypothesis on the orientation characters is often redundant.

Theorem 7. *Let $f : M \to N$ be a 2-connected map between finite PD_4-complexes with $\chi(M) = \chi(N)$. If $H^2(N; Z/2Z) \neq 0$ then $f^*w_1(N) = w_1(M)$, and if moreover N is orientable and $H^2(N; Q) \neq 0$ then f is a homotopy equivalence.*

Proof. Since f is 2-connected $H^2(f; Z/2Z)$ is injective, and since $\chi(M) = \chi(N)$ it is an isomorphism. Since $H^2(N; Z/2Z) \neq 0$, the nondegeneracy of Poincaré duality implies that $H^4(f; Z/2Z) \neq 0$, and so f is a $Z/2Z$-(co)homology equivalence. Since $w_1(M)$ is characterized by the Wu formula $x \cup w_1(M) = Sq^1 x$ for all x in $H^3(M; Z/2Z)$, it follows that $f^*w_1(N) = w_1(M)$.

If $H^2(N; Q) \neq 0$ then $H^2(N; Z)$ has positive rank and $H^2(N; Z/2Z) \neq 0$, so N orientable implies M orientable. We may then repeat the above argument with integral coefficients, to conclude that f has degree ± 1. The result then follows from Theorem 3. //

The argument breaks down if, for instance, $M = S^1 \tilde{\times} S^3$ is the nonorientable S^3-bundle over S^1, $N = S^1 \times S^3$ and f is the composite of the projection of M onto S^1 followed by the inclusion of a factor.

Corollary. *Let M be a finite PD_4-complex with fundamental group π. If π is a PD_4-group of type FF, $H_2(\pi; Z/2Z) \neq 0$ and $\chi(M) = \chi(\pi)$ then M is aspherical.* //

We would like to replace the hypothesis in Theorem 3 that there be a map f as above by weaker, more algebraic conditions. If M and N are closed 4-manifolds with isomorphic algebraic 2-types then there is a 3-connected map $f: M \to P_2(N)$. The restriction of such a map to $M_o = M\backslash D^4$ is homotopic to a map $f_o: M_o \to N$ which induces isomorphisms on π_i for $i \leq 2$. In particular, $\chi(M) = \chi(N)$. Thus if f_o extends to a map from M to N we may be able to apply Theorem 7. However we usually need more information on how the top cell is attached. The characteristic classes and the equivariant intersection pairing on $\pi_2(M)$ are the obvious candidates. (See Chapter IX).

The following criterion arises in studying the homotopy types of circle bundles over 3-manifolds. (See Chapter III).

Theorem 8. *Let E be a finite PD_4-complex with fundamental group π and suppose that $H^4(f_E; Z^{w_1(E)})$ is a monomorphism. A finite PD_4-complex M is homotopy equivalent to E if and only if there is an isomorphism θ from $\pi_1(M)$ to π such that $w_1(M) = w_1(E)\theta$, there is a lift $\hat{c}: M \to P_2(E)$ of θc_M such that $\hat{c}_*[M] = \pm f_{E*}[E]$ and $\chi(M) = \chi(E)$.*

Proof. The conditions are clearly necessary. Conversely, suppose that they hold. We shall adapt to our situation the arguments of Hendriks [He77] in analyzing the obstructions to the existence of a degree 1 map between PD_3-complexes realizing a given homomorphism of fundamental groups. For simplicity of notation we shall write $\tilde{Z}$ for $Z^{w_1(E)}$ and also for $Z^{w_1(M)} (= \theta^*\tilde{Z})$, and use θ to identify $\pi_1(M)$ with π and $K(\pi_1(M), 1)$ with $K(\pi, 1)$. We may suppose the sign of the fundamental class $[M]$ is so chosen that $\hat{c}_*[M] = f_{E*}[E]$.

Let $E_o = E\backslash D^4$. Then $P_2(E_o) = P_2(E)$ and may be constructed as the union of E_o with cells of dimension ≥ 4. Let $h: \tilde{Z} \otimes_{Z[\pi]} \pi_4(P_2(E_o), E_o) \to H_4(P_2(E_o), E_o; \tilde{Z})$ be the $w_1(E)$-twisted relative Hurewicz homomorphism, and let ∂ be the connecting homomorphism from $\pi_4(P_2(E_o), E_o)$ to $\pi_3(E_o)$ in the exact sequence of homotopy for the pair $(P_2(E_o), E_o)$. Then h and ∂ are isomorphisms since f_{E_o} is 3-connected. The composite of the inclusion $H_4(P_2(E); \tilde{Z}) = H_4(P_2(E_o); \tilde{Z}) \to H_4(P_2(E_o), E_o; \tilde{Z})$ with h^{-1} and $1 \otimes_\mu \partial$ gives a monomorphism τ_E from $H_4(P_2(E); \tilde{Z})$ to $\tilde{Z} \otimes_{Z[\pi]} \pi_3(E_o)$. Similarly

$M_o = M \backslash D^4$ may be viewed as a subspace of $P_2(M_o)$ and there is a monomorphism τ_M from $H_4(P_2(M);\tilde{Z})$ to $\tilde{Z}\otimes_{Z[\pi]}\pi_3(M_o)$. These monomorphisms are natural with respect to maps defined on the 3-skeleta of the spaces (i.e., E_o and M_o).

The classes $\tau_E(f_{E*}[E])$ and $\tau_M(f_{M*}[M])$ are the images of the primary obstructions to retracting E onto E_o and M onto M_o, under the Poincaré duality isomorphisms from $H^4(E,E_o;\pi_3(E_o))$ to $H_0(E\backslash E_o;\tilde{Z}\otimes_{Z[\pi]}\pi_3(E_o)) = \tilde{Z}\otimes_{Z[\pi]}\pi_3(E_o)$ and $H^4(M,M_o;\pi_3(M_o))$ to $\tilde{Z}\otimes_{Z[\pi]}\pi_3(M_o)$, respectively. Since M_o is homotopy equivalent to a cell complex of dimension ≤ 3 the restriction of $\hat{c}$ to M_o is homotopic to a map from M_o to E_o. In particular, $(1\otimes_{Z[\pi]}\hat{c}_\sharp)\tau_M(f_{M*}[M]) = \tau_E(f_{E*}[E])$, where $\hat{c}_\sharp$ is the homomorphism from $\pi_3(M_o)$ to $\pi_3(E_o)$ induced by $\hat{c}|M_o$. It follows as in [He77] that the obstruction to extending $\hat{c}|M_o : M_o \to E_o$ to a map d from M to E is trivial.

Since $f_{E*}d_*[M] = \hat{c}_*[M] = f_{E*}[E]$ and f_{E*} is a monomorphism in degree 4 the map d has degree 1. Thus d is a homotopy equivalence, by Theorem 3. //

If there is such a lift $\hat{c}$ then $c_M^*\theta^*k_1(E) = 0$ and $\theta_*c_{M*}[M] = c_{E*}[E]$.

3. Finitely dominated covering spaces.

In this section we shall show that if a PD_4-complex has an infinite regular covering space which is finitely dominated then either the complex is aspherical or its universal covering space is homotopy equivalent to S^2 or S^3. In Chapters III and IV we shall see that such manifolds are close to being total spaces of fibre bundles.

Theorem 9. *Let M be a PD_4-complex with fundamental group π. Suppose that $p:\hat{M}\to M$ is a regular covering map, with covering group $G = Aut(p)$, and such that $\hat{M}$ is finitely dominated. Then*

(i) G has finitely many ends;

(ii) if $\hat{M}$ is acyclic then it is contractible and M is aspherical;

(iii) if G has one end and $\pi_1(\hat{M})$ is infinite and FP_3 then M is aspherical and $\hat{M}$ is homotopy equivalent to S^1 or to an aspherical closed surface;

(iv) if G has one end and $\pi_1(\hat{M})$ is finite but $\hat{M}$ is not acyclic then $\hat{M}\simeq S^2$

or RP^2;

(v) G *has two ends if and only if* $\hat{M}$ *is a* PD_3*-complex.*

Proof. We may clearly assume that G is infinite and that M is orientable. As $Z[G]$ has no nonzero left ideal (i.e., submodule) which is finitely generated as an abelian group $Hom_{Z[G]}(H_p(\hat{M};Z),Z[G]) = 0$ for all $p \geq 0$, and so the bottom row of the UCSS for the covering p is 0. From Poincaré duality and the UCSS we find that $H^1(G;Z[G]) \cong \overline{H_3(\hat{M};Z)}$. As this group is finitely generated, and as G is infinite, G has one or two ends.

If $\hat{M}$ is acyclic then G is a PD_4-group and so $\hat{M}$ is a PD_0-complex, hence contractible, by [Go79]. Hence M is aspherical.

Suppose that G has one end. Then $H_3(\hat{M};Z) = H_4(\hat{M};Z) = 0$. Since $\hat{M}$ is finitely dominated the chain complex $C_*(\hat{M})$ is chain homotopy equivalent over $Z[\pi_1(\hat{M})]$ to a complex D_* of finitely generated projective $Z[\pi_1(\hat{M})]$-modules. If $\pi_1(\hat{M})$ is FP_3 then the aumentation $Z[\pi_1(\hat{M})]$-module Z has a free resolution P_* which is finitely generated in degrees ≤ 3. On applying Schanuel's Lemma to the exact sequences $0 \to Z_2 \to D_2 \to D_1 \to D_0 \to Z \to 0$ and $0 \to \partial P_3 \to P_2 \to P_1 \to P_0 \to Z \to 0$ derived from these two chain complexes we find that Z_2 is finitely generated as a $Z[\pi_1(\hat{M})]$-module. Hence $\Pi = \pi_2(M) = \pi_2(\hat{M})$ is also finitely generated as a $Z[\pi_1(\hat{M})]$-module and so $Hom_\pi(\Pi, Z[\pi]) = 0$. If moreover $\pi_1(\hat{M})$ is infinite then $H^s(\pi;Z[\pi]) = 0$ for $s \leq 2$, so $\Pi = 0$, by Lemma 4, and M is aspherical. We may then apply an LHSSS corner argument to show that if $H^2(G;Z[G]) \neq 0$ then $\pi_1(\hat{M})$ has one end and $H^2(\pi_1(\hat{M});Z[\pi_1(\hat{M})]) \cong Z$. Since $\pi_1(\hat{M})$ is FP, it is a PD_2-group, by Theorem 3 of [Fa75], and so $\hat{M}$ is homotopy equivalent to an aspherical closed surface, by [EM80] and [EL83]. If $H^2(G;Z[G]) = 0$ then $H^1(\pi_1(\hat{M});Z[\pi_1(\hat{M})]) \cong Z$, by another LHSSS corner argument. Since $\pi_1(\hat{M})$ is FP and hence torsion free, it must be infinite cyclic. Therefore $\hat{M} \simeq S^1$.

If $\pi_1(\hat{M})$ is finite but $\hat{M}$ is not acyclic then the universal covering space $\tilde{M}$ is also finitely dominated but not contractible, and $\Pi = H_2(\tilde{M};Z)$ is a nontrivial finitely generated abelian group, while $H_3(\tilde{M};Z) = H_4(\tilde{M};Z) = 0$. Moreover $H^2(\pi;Z[\pi]) \cong \bar{\Pi}$, by Lemma 4. Let C be a finite cyclic subgroup of π which acts trivially on Π. Then it follows easily from the Cartan-Leray spectral sequence for the projection of $\tilde{M}$ onto $\tilde{M}/C$ that there are isomorphisms

$H_{n+3}(C;Z) \cong H_n(C;\Pi)$, for all $n > 2$. If n is odd this isomorphism reduces to $0 = \Pi/|C|\Pi$. Since Π is finitely generated, this implies that multiplication by $|C|$ is an isomorphism. On the other hand, if n is even then we have $Z/|C|Z \cong \{a \in \Pi : |C|a = 0\}$. Hence we must have $C = 1$. Now since Π is finitely generated, any torsion subgroup of $Aut(\Pi)$ is finite. (Let T be the torsion subgroup of Π and suppose that $\Pi/T \cong Z^r$. Then the natural homomorphism from $Aut(\Pi)$ to $Aut(\Pi/T)$ has finite kernel, and its image is isomorphic to a subgroup of $GL(r,Z)$, which is virtually torsion free). Hence as π is infinite it must have elements of infinite order, and so $H^2(\pi;Z[\pi])$ must be infinite cyclic, by Corollary 5.2 of [Fa74], as it is a nontrivial finitely generated group. Hence $\tilde{M} \simeq S^2$ and $\pi_1(\hat{M})$ has order at most 2, so $\hat{M} \simeq S^2$ or RP^2.

Suppose now that G has two ends. We may assume that $G \cong Z$ and it then follows easily from [Go79] that $\hat{M}$ is a PD_3-complex. Conversely, if $\hat{M}$ is a PD_3-complex then G is infinite and $H^1(G;Z[G]) \cong H_3(\hat{M};Z)$, and so G has two ends. //

Is the hypothesis in *(iii)* that $\pi_1(\hat{M})$ be FP_3 redundant?

Corollary A. *The covering space $\hat{M}$ is homotopy equivalent to a closed surface if and only if it is finitely dominated, $H^2(G;Z[G]) \cong Z$ and $\pi_1(\hat{M})$ is FP_3. The group G is then FP_∞.*

Proof. If $\hat{M}$ is homotopy equivalent to a closed surface then it is finitely dominated and $\pi_1(\hat{M})$ is FP_3, and the conditions on G in the first assertion follow from the UCSS. The converse follows from parts *(iii)* and *(iv)* of the theorem, with Theorem I.11.

If $\tilde{M} \simeq S^2$ and π is FP_n we may splice together a cellular chain complex for $\tilde{M}$ and a finite partial resolution of the augmentation module Z to show that π is FP_{n+3}. (Since M is homotopy equivalent to a finite cell complex the singular chain complex of $\tilde{M}$ is chain homotopy equivalent to a finite free $Z[\pi]$-complex C_*. Let P_* be a projective resolution of the augmentation module Z in which $P_0 = Z[\pi]$ and P_i is finitely generated for all $i \leq n$. Let $Q_n = C_n$ if $n \leq 2$ and $Q_n = C_n \oplus P_{n-3}$ if $n \geq 3$, where the differential from Q_3 to Q_2 maps a generator of the summand P_0 onto a 2-cycle representing the generator of $H_2(\tilde{M};Z)$). An induction on n then shows that π is FP_∞. If M is aspherical then π is a PD_4-group and so is FP. Since finite groups are FP_∞ and surface groups are FP it follows that π and $\pi_1(M_G)$ are both FP_∞, in all cases. Hence so is $G \cong \pi/\pi_1(\hat{M})$, by Proposition 2.7 of [Bi]. //

Corollary B. *The covering space $\hat{M}$ is homotopy equivalent to S^1 if and only if it is finitely dominated, G has one end, $H^2(G;Z[G]) = 0$ and $\pi_1(\hat{M})$ is a nontrivial finitely generated free group.*

Proof. If $\hat{M} \simeq S^1$ then it is finitely dominated and M is aspherical, and the conditions on G follow from the LHSSS. The converse follows from part *(iii)* of the theorem, since a nontrivial finitely generated free group is infinite and FP. //

In the simply connected case "finitely dominated" is equivalent to "homotopy equivalent to a finite complex".

Corollary C. *If $H_*(\tilde{M};Z)$ is finitely generated then either M is aspherical or $\tilde{M}$ is homotopy equivalent to S^2 or S^3 or $\pi_1(M)$ is finite.* //

We shall examine the spherical cases more closely in the the next section.

4. Spherical universal covering spaces

The determination of the PD_4-complexes with universal covering space homotopy equivalent to S^3 rests upon the known structure of groups G with $H^1(G;Z[G]) \cong Z$.

Theorem 10. *Let M be a PD_4-complex with fundamental group π. Then*
(i) $\tilde{M} \simeq S^3$ if and only if π has two ends and $\chi(M) = 0$; if so
(ii) the group π has a maximal finite normal subgroup F, which has cohomological period dividing 4, and the corresponding covering space M_F has the homotopy type of an orientable PD_3-complex;
(iii) the homotopy type of M is determined by π and $k_2(M) \in H^4(\pi;\pi_3(M))$, the first nontrivial k-invariant of M, which restricts to a generator of $H^4(F;\pi_3(M)) = H^4(F;Z)$, modulo the action of $Aut(\pi)$.

Proof. If $\tilde{M} \simeq S^3$ then $H^1(\pi; Z[\pi]) \cong Z$ and so π has two ends. Hence π is virtually Z. Let M_Z be an orientable finite covering space corresponding to an infinite cyclic subgroup. Then M_Z is homotopy equivalent to the mapping torus of a self homotopy equivalence of $S^3 \simeq \tilde{M}$, so $\chi(M_Z) = 0$ and hence $\chi(M) = 0$ also.

Suppose conversely that $\chi(M) = 0$ and π is virtually Z. Let M_Z be an orientable finite covering space with fundamental group Z. Then $\chi(M_Z) = 0$ and so $H_2(M_Z; Z) = 0$. The homology groups of $\tilde{M} = \tilde{M}_Z$ may be regarded as modules over $Z[t, t^{-1}] \cong Z[Z]$. By the Wang sequence for the projection of $\tilde{M}$ onto M_Z, multiplication by $t - 1$ maps $H_2(\tilde{M}; Z)$ onto itself. Therefore $Hom_{Z[\pi]}(H_2(\tilde{M}; Z), Z[\pi]) = 0$. Since π has two ends $H^2(\pi; Z[\pi]) = 0$ and so $H_2(\tilde{M}; Z) = 0$, by Lemma 4, while $H_3(\tilde{M}; Z) \cong Z$ and $H_4(\tilde{M}; Z) = 0$. Therefore the map from S^3 to $\tilde{M}$ representing a generator of $\pi_3(M)$ is a homotopy equivalence.

Let F be the maximal finite normal subgroup of π. Since F acts freely on $\tilde{M} \simeq S^3$ it has cohomological period dividing 4 and $M_F = \tilde{M}/F$ is a PD_3-complex. Suppose that M_F is nonorientable, and let C be a cyclic subgroup of F generated by an orientation reversing element. The classifying map from $M_C = \tilde{M}/C$ to $K(C, 1)$ is 3-connected and so there are isomorphisms $H_1(M_C; Z) \cong H^2(M_C; \tilde{Z}) \cong H^2(C; \tilde{Z}) = 0$, where $\tilde{Z}$ is the nontrivial infinite cyclic $Z[C]$-module. But this contradicts $H_1(M_C; Z) \cong C \neq 0$. Thus M_F must be orientable.

Since $\pi_2(M) = 0$ and π acts on $\pi_3(M) \cong Z$ via the orientation character the first nonzero k-invariant lies in $H^4(\pi; Z^w)$. The image in $H^4(F; Z^w)$ is the k-invariant for the PD_3-complex M_F, and so generates this group [Wl67]. Let $P_3(M)$ be the third stage of the Postnikov tower for M and let $j : M \to P_3(M)$ be the natural map. Then j is 4-connected and we may construct $P_3(M)$ by adjoining cells of dimension greater than 4 to M. If M_1 is another such finite PD_4-complex and $\theta : \pi_1(M_1) \to \pi$ is an isomorphism which identifies the k-invariants then there is a 4-connected map $j_1 : M_1 \to P_3(M)$ inducing θ, which is homotopic to a map with image in the 4-skeleton of $P_3(M)$, and so there is a map $h : M_1 \to M$ such that j_1 is homotopic to jh. The map h induces isomorphisms on π_i for $i \leq 3$, since j and j_1 are 4-connected, and so the lift $\tilde{h} : \tilde{M}_1 \simeq S^3 \to \tilde{M} \simeq S^3$ is a homotopy equivalence, by the theorems of Hurewicz and Whitehead. Thus h is itself a homotopy equivalence. //

Every finite group F with cohomological period dividing 4 is the fundamental group of some orientable PD_3-complex X [Sw60]; the space $X \times S^1$ is then a finite PD_4-complex with fundamental group $F \times Z$ and Euler characteristic 0. In Chapter VIII we shall show that the universal covering spaces of 4-manifolds as in Theorem 10 are homeomorphic to $S^3 \times R$, and we shall consider further the possible fundamental groups of such complexes and 4-manifolds.

The determination of the PD_4-complexes with universal covering space homotopy equivalent to S^2 rests on Farrell's structure theorem for groups G with $H^2(G; Z[G]) \cong Z$.

Theorem 11. *Let M be a PD_4-complex with fundamental group π. Then*
(i) the universal covering space $\tilde{M}$ is homotopy equivalent to S^2 if and only if π is infinite and $\pi_2(M) \cong Z$;
(ii) if these conditions hold then the kernel κ of the natural homomorphism u from π to $Aut(\pi_2(M)) = Z/2Z$ is torsion free and $H^2(\kappa; Z[\kappa]) \cong Z$;
(iii) if moreover π has a subgroup of finite index with infinite abelianization then κ is the fundamental group of an aspherical closed surface.

Proof. The conditions are clearly necessary. Suppose that they hold. On passing to a finite covering if necessary we may assume that M is orientable and that π acts trivially on $\pi_2(M)$. Then $H_0(\tilde{M}; Z) = Z$, $H_1(\tilde{M}; Z) = 0$ and $H_2(\tilde{M}; Z) = \pi_2(M)$, while $H_3(\tilde{M}; Z) \cong \bar{H}^1(\pi; Z[\pi])$ and $H_4(\tilde{M}; Z) = 0$. Now $Hom_{Z[\pi]}(\pi_2(M), Z[\pi]) = 0$, since π is infinite and $\pi_2(M) \cong Z$. Therefore $H^2(\pi; Z[\pi])$ is infinite cyclic, by Lemma 4, and so π has one end, by Theorem I.11. Hence $H_3(\tilde{M}; Z) = 0$ and so $\tilde{M} \simeq S^2$.

Now assume that $\tilde{M} \simeq S^2$. Note that if ν is any subgroup of finite index in π then ν has one end and $H^2(\nu; R[\nu]) \cong H_2(\tilde{M}; R) \cong R$, for $R = Z$ or a field. Suppose that g is a nontrivial element of finite order in π, and let G be the subgroup generated by g. The Gysin sequence of the fibration $\tilde{M} \to \tilde{M}/G \to K(G, 1)$ gives isomorphisms from $H_{n+3}(G; Z)$ to $H_n(G; H_2(\tilde{M}; Z))$

for all $n \geq 2$, since $H_s(\tilde{M}/G; Z) = 0$ for all $s \geq 4$. Therefore g acts nontrivially on $\pi_2(M) = H_2(\tilde{M}; Z)$ and so $\kappa = \mathrm{Ker} u$ is torsion free.

Suppose now that π has a subgroup of finite index with infinite abelianization, and let ρ be the intersection of this subgroup with κ. Then ρ also has finite index in π and has infinite abelianization. Therefore it is an HNN extension $\rho \cong H*_J$ where the base H and associated subgroups J and tJt^{-1} are finitely generated, by Theorem A of [BS78]. Since ρ is torsion free and $H^2(\rho; Z/2Z[\rho])$ is 1-dimensional any finitely generated subgroup of infinite index in ρ is free, by Theorem 2.2 of [Fa74]. Therefore H and J are free and so ρ has cohomological dimension at most 2. Since it is finitely presentable, has one end and $H^2(\rho; Z[\rho]) \cong Z$ it is a PD_2-group [Bi]. Since κ is torsion free it is also a PD_2-group and so is a surface group [EM80]. //

If π is finite and $\pi_2(M) \cong Z$ then it is not hard to see that $\tilde{M} \simeq CP^2$ and hence $\pi = 1$. (See Chapter IX). In Chapter VII we shall show that the universal covering spaces of 4-manifolds as in Theorem 11.(*iii*) are homeomorphic to $S^2 \times R^2$.

The reference to [BS78] could be replaced by the observation that duality and the spectral sequence imply that $H^3(\rho; W) = 0$ for any free $Z[\rho]$-module W and reference to Corollary 2.3 of [Fa74]. The condition "π has a subgroup of finite index with infinite abelianization" could be replaced by "$v.c.d.\pi < \infty$". Can it be dispensed with entirely? In other words, if $\tilde{M} \simeq S^2$ and π acts trivially on $\pi_2(M)$ is $H^1(M; Z) \neq 0$?

The arguments in this section extend to the study of PD_n-complexes with universal covering space homotopy equivalent to S^{n-1} or S^{n-2}. The higher codimension cases appear to be less accessible; see [St92] for a survey of aspects of this "mixed spaceform" problem.

5. Minimizing the Euler characteristic

It is well known that every finitely presentable group is the fundamental group of some closed orientable 4-manifold. Such manifolds are far from unique, for the Euler characteristic may be made arbitrarily large by taking connected sums with simply connected manifolds. Following Hausmann and

Weinberger we may define an invariant $q(\pi)$ for any finitely presentable group π by $q(\pi) = \min\{\chi(M) | M \textit{ is a } PD_4\textit{-complex with } \pi_1(M) \cong \pi\}$ [HW85]. We may also define related invariants q^X where the minimum is taken over the class of PD_4-complexes whose normal fibration has an X-reduction. An elementary argument shows that $q^{SG}(\pi) \geq 2 - 2\beta_1(\pi) + \beta_2(\pi)$, where SG is the class of orientable PD_4-complexes. If π is a finitely presentable, orientable PD_4-group of type FF we see immediately that $q^{SG}(\pi) \geq \chi(\pi)$. Since Euler characteristics are multiplicative with respect to passage to finite covers it follows easily that $q(\pi) = \chi(\pi)$ if $K(\pi, 1)$ is a finite PD_4-complex.

Theorem 12. *Let M be a finite PD_4-complex with fundamental group π. Suppose that $c.d._Q\pi \leq 2$ and that the augmentation $Q[\pi]$-module has a finitely generated free resolution. Then $\chi(M) \geq 2\chi(\pi; Q) = 2(1-\beta_1(\pi; Q)+\beta_2(\pi; Q))$. If moreover $\chi(M) = 2\chi(\pi; Q)$ then $\pi_2(M) \cong \overline{H^2(\pi; Z[\pi])}$.*

Proof. Since $Q[\pi]$ has global dimension 2, we may assume without loss of generality that there is an exact sequence

$$0 \to Q[\pi]^r \to Q[\pi]^g \to Q[\pi] \to Q \to 0. \tag{1}$$

We may also assume that M is a finite 4-dimensional cell complex. Let C_* be the cellular chain complex of the universal covering space $\tilde{M}$, with coefficients Q, and let $H_i = H_i(C_*) = H_i(\tilde{M}; Q)$ and $H^i = H^i(Hom_{Q[\pi]}(C_*, Q[\pi]))$. Since $\tilde{M}$ is simply connected and π is infinite, $H_0 \cong Q$ and $H_1 = H_4 = 0$. The chain complex C_* breaks up into exact sequences:

$$0 \to C_4 \to Z_3 \to H_3 \to 0, \tag{2}$$

$$0 \to Z_3 \to C_3 \to Z_2 \to H_2 \to 0, \tag{3}$$

$$0 \to Z_2 \to C_2 \to C_1 \to C_0 \to Q \to 0. \tag{4}$$

We shall let $e^i N = Ext^i_{Q[\pi]}(N, Q[\pi])$, to simplify the notation in what follows. The rational analogue of Lemma 4 gives another exact sequence:

$$0 \to e^2 Q \to H^2 \to e^0 H_2 \to 0 \tag{5}$$

and isomorphisms $H^1 \cong e^1 Q$ and $e^1 H_2 = e^2 H_3 = 0$. Poincaré duality gives further isomorphisms $H^1 \cong \overline{H_3}$, $H^2 \cong \overline{H_2}$, $H^3 = 0$ and $H^4 \cong \bar{Q}$.

Applying Schanuel's Lemma to the sequences (1) and (4) we obtain $Z_2 \oplus C_1 \oplus Q[\pi] \oplus Q[\pi]^r \cong C_2 \oplus C_0 \oplus Q[\pi]^g$, so Z_2 is a finitely generated stably free module. Similarly, Z_3 is projective, since $p.d.H_2 \leq 2 = gl.dim.Q[\pi]$. Since π is finitely presentable it is accessible, and hence $e^1 Q$ is finitely generated as a $Q[\pi]$-module, by Theorems IV.7.5 and VI.6.3 of [DD]. Therefore Z_3 is also finitely generated, since it is an extension of $H_3 \cong \bar{e}^1 Q$ by C_4. Dualizing the sequence (3) and using the fact that $e^1 H_2 = 0$ we obtain an exact sequence of right modules

(6) $$0 \to e^0 H_2 \to e^0 Z_2 \to e^0 C_3 \to e^0 Z_3 \to e^2 H_2 \to 0.$$

Since duals of finitely generated projective modules are projective, it follows that $e^0 H_2$ is projective. Hence the sequence (5) gives $H^2 \cong e^0 H_2 \oplus e^2 Q$.

Let I be the augmentation ideal of $Q[\pi]$. Then there are exact sequences

(7) $$0 \to Q[\pi]^r \to Q[\pi]^g \to I \to 0$$

and

(8) $$0 \to I \to Q[\pi] \to Q \to 0.$$

Dualizing, we obtain exact sequences of right modules

(9) $$0 \to e^0 I \to Q[\pi]^g \to Q[\pi]^r \to e^2 Q \to 0$$

and

(10) $$0 \to Q[\pi] \to e^0 I \to e^1 Q \to 0.$$

Applying Schanuel's Lemma twice more, to the pairs of sequences (2) and the conjugate of (10) (using $H_3 \cong \overline{e^1 Q}$) and to (3) and the conjugate of (9) (using $H_2 \cong \overline{e^0 H_2 \oplus e^2 Q}$) and putting all together, we obtain an isomorphism $Z_3 \oplus (Q[\pi]^{2g} \oplus C_0 \oplus C_2 \oplus C_4) \cong Z_3 \oplus (Q[\pi]^{2+2r} \oplus C_1 \oplus C_3 \oplus \overline{e^0 H_2})$. Since all the summands are finitely generated projective modules, it follows that

$\overline{e^0 H_2}$ is stably free. On adding a projective complement for Z_3 to each side of this equation and comparing ranks we find that the rank of $\overline{e^0 H_2}$ is $\chi(M) - 2(1 - g + r)$. As $Z[\pi]$ has the SIBN property this must be nonnegative and as we see immediately from sequence (1) that $\chi(\pi; Q) = 1 - g + r$ we have $\chi(M) \geq 2\chi(\pi; Q)$.

If $\chi(M) = 2\chi(\pi; Q)$ then $e^0 H_2 = 0$ and so $Hom_{Z[\pi]}(H_2(\tilde{M}; Z), Z[\pi]) = 0$. As $\pi_2(M) \cong H_2(\tilde{M}; Z)$ the final assertion now follows from the integral analogue of sequence (5). //

If $H_2(\pi; Q) \neq 0$ the theorem gives a better estimate for $q(\pi)$ than the general estimate given above.

Corollary. *If $\pi = \pi_1(P)$ where P is an aspherical finite 2-complex then $q(\pi) = 2\chi(P)$, and the minimum is realized by an s-parallelizable PL 4-manifold. If moreover π has one end and M is any closed 4-manifold with $\pi_1(M) \cong \pi$ then $\pi_2(M)$ is stably isomorphic to $\bar{H}^2(\pi; Z[\pi])$ and the first k-invariant of M is 0.*

Proof. If we choose a PL embedding $j : P \to R^5$, the boundary of a regular neighbourhood N of $j(P)$ is an s-parallelizable PL 4-manifold with fundamental group π and with Euler characteristic $2\chi(P)$. By the theorem, this is best possible.

The second assertion follows on reworking the argument of the theorem with $Z[\pi]$ coefficients. Since $c.d.\pi \leq 2$ we have $H^3(\pi; \pi_2(M)) = 0$ and $k_1(M) = 0$. //

Note that the conjugation of the module structure involves the orientation character of M. Is the assumption that π has one end needed for the second assertion? By Theorem II.6 of [H] a finitely presentable group is the fundamental group of an aspherical finite 2-complex if and only if it has cohomological dimension ≤ 2 and is efficient, i.e. has a presentation of deficiency $\beta_1(\pi; Q) - \beta_2(\pi; Q)$. It is not known whether every finitely presentable group of cohomological dimension 2 is efficient.

Theorem 13. *Let N be a PD_3-complex with torsion free fundamental group ν. Then*

(i) c.d.$\pi \leq 3$;

(ii) the $Z[\nu]$-module $\pi_2(N)$ is finitely presentable and has projective dimension at most 1;

(iii) if ν is not a free group then $\pi_2(N)$ is projective;

(iv) if ν is not a free group then any two of the conditions "ν is FF", "N is homotopy equivalent to a finite complex" and "$\pi_2(N)$ is stably free" imply the third.

Proof. We may clearly assume that $\nu \neq 1$. The PD_3-complex N is homotopy equivalent to a connected sum of aspherical PD_3-complexes and a 3-manifold with free fundamental group, by Theorem 2 of [Tu90]. Therefore ν is a corresponding free product, and so it has cohomological dimension at most 3 and is FP. Since N is finitely dominated the equivariant chain complex of the universal covering space $\tilde{N}$ is chain homotopy equivalent to a complex $0 \to C_3 \to C_2 \to C_1 \to C_0 \to 0$ of finitely generated projective left $Z[\nu]$-modules. Then the sequences $0 \to Z_2 \to C_2 \to C_1 \to C_0 \to Z \to 0$ and $0 \to C_3 \to Z_2 \to \pi_2(N) \to 0$ are exact, where Z_2 is the module of 2-cycles in C_2. Since ν is FP and $c.d.\nu \leq 3$ Schanuel's lemma implies that Z_2 is projective and finitely generated. Hence $\pi_2(N)$ has projective dimension at most 1, and is finitely presentable.

It follows easily from the UCSS and Poincaré duality that $\pi_2(N)$ is isomorphic to $\overline{H^1(\nu; Z[\nu])}$ and that there is an exact sequence

$$H^3(\nu; Z[\nu]) \to H^3(N; Z[\nu]) \to Ext^1_\nu(\pi_2(N), Z[\nu]) \to 0.$$

The $w_1(N)$-twisted augmentation homomorphism from $Z[\nu]$ to $\tilde{Z}$ which sends $g \in \nu$ to $w_1(N)(g)$ induces an isomorphism from $H^3(N; Z[\nu])$ to $H^3(N; \tilde{Z}) \cong Z$. If ν is free the first term in this sequence is 0, and so $Ext^1_\nu(\pi_2(N), Z[\nu]) \cong Z$. (In particular, $\pi_2(N)$ has projective dimension 1). If ν is *not* free then the map $H^3(\nu; Z[\nu]) \to H^3(N; Z[\nu])$ in the above sequence is onto, as can be seen by comparison with the corresponding sequence with coefficients $\tilde{Z}$. Therefore $Ext^1_\nu(\pi_2(N), Z[\nu]) = 0$. Since $\pi_2(N)$ has a short resolution by finitely generated projective modules, it follows that it is in fact projective.

The final assertion follows easily from the fact that if $\pi_2(N)$ is projective then $Z_2 \cong \pi_2(N) \oplus C_3$. //

Swarup showed that if N is a 3-manifold which is the connected sum of a 3-manifold whose fundamental group is free of rank r with $s \geq 1$ aspherical 3-manifolds then $\pi_2(N)$ is a finitely generated free $Z[\nu]$-module of rank $2r+s-1$ [Sw73]. If ν is not torsion free then the projective dimension of $\pi_2(N)$ is infinite.

The following simple result will be invoked in several of the later chapters.

Lemma 14. *If M is an orientable PD_4-complex and $\chi(M) \leq 0$ then the abelianization $\pi_1(M)/\pi_1(M)' = H_1(M; Z)$ is infinite.*

Proof. Since M is connected and orientable $\beta_4(M) = \beta_0(M) = 1$ and $\beta_3(M) = \beta_1(M)$, so $\chi(M) = 2 - 2\beta_1(M) + \beta_2(M)$. Therefore $\chi(M) \leq 0$ implies $\beta_1(M) \geq 1$. //

The assumption of orientability is necessary: if $M = RP^4 \sharp RP^4$ then $\chi(M) = 0$ and $\pi_1(M) \cong D$, so $\pi_1(M)/\pi_1(M)' \cong (Z/2Z)^2$.

CHAPTER III

MAPPING TORI AND CIRCLE BUNDLES

The study of 4-manifolds which fibre over the circle benefits from both 4-dimensional surgery and from 3-manifold theory, which may be applied to the fibre and the characteristic map. Unfortunately, there is as yet no satisfactory fibration theorem in this dimension. We shall give a homotopy approximation to such a theorem. By a result of Quinn, an infinite cyclic covering space of a closed n-manifold M is homotopy equivalent to a PD_{n-1}-complex if and only if it is finitely dominated. The fundamental group ν of the covering space must then be finitely generated, and $\chi(M) = 0$. It is conceivable that these conditions may suffice when $n = 4$; we shall see that this is so if also either ν is free or ν does not contain the centre of $\pi_1(M)$.

In the final section we give conditions for a 4-manifold to be homotopy equivalent to the total space of an S^1-bundle over a PD_3-complex, and show that these conditions are sufficient if the fundamental group of the PD_3-complex is torsion free but not free.

1. A general criterion

Let E be a connected cell complex and let $f : E \to S^1$ be a map which induces an epimorphism f_* from $\pi_1(E)$ to $Z = \pi_1(S^1)$ with kernel ν. We may identify the homotopy fibre of f with $E_\nu = E \times_{S^1} R = \{(x, y) \in E \times R | f(x) = e^{2\pi iy}\}$, the covering space of E with group ν. The covering group is generated by $\phi : E_\nu \to E_\nu$ where $\phi(x, y) = (x, y + 1)$ for all (x, y) in E_ν, and we can recover E up to homotopy as the mapping torus of ϕ.

Quinn showed that the total space of a fibration with finitely dominated base and fibre is a Poincaré duality complex if and only if both the base and fibre are Poincaré duality complexes. (Quinn did not publish his argument, but see [Go79] for a very elegant proof). In particular, if E is a PD_4-complex

Typeset by $\mathcal{A}_\mathcal{M}\mathcal{S}$-TEX

and E_ν is finitely dominated then it is in fact a PD_3-complex. We would like to deduce this conclusion from hypotheses on the fundamental group and Euler characteristic alone. If E_ν is finitely dominated then ν is finitely presentable, and $H_*(E_\nu; Z)$ is finitely generated. It then follows from the Wang sequence for the projection of E_ν onto E that $\chi(E) = 0$. In our first result we shall show that such mapping tori have minimal Euler characteristic for their fundamental groups (i.e. $\chi(M) = q(\pi)$). In Theorem 2 we shall use the algebraic characterization of 3-complexes, while in Theorem 3 we shall show directly that E_ν satisfies Poincaré duality with local coefficients.

Theorem 1. *Let M be a closed 4-manifold such that $\pi = \pi_1(M)$ is an extension of Z by a finitely generated normal subgroup ν. Then $\chi(M) \geq 0$, and hence $q(\pi) \geq 0$.*

Proof. Since Euler characteristics are multiplicative in finite coverings, we may assume that M is orientable. Let $Q\Lambda = Q[t, t^{-1}]$ be the rational group ring of Z, and let M' be the infinite cyclic covering space of M with fundamental group ν. Since M is compact and $Q\Lambda$ is noetherian the homology groups $H_i(M'; Q) = H_i(M; Q\Lambda)$ are finitely generated as $Q\Lambda$-modules. Moreover as ν is finitely generated π has one or two ends and these homology groups are finite dimensional as Q-vector spaces, except perhaps when $i = 2$. Now $H_2(M'; Q) \cong Q^r \oplus Q\Lambda^s$ for some $r, s \geq 0$, by the Structure Theorem for modules over a P.I.D. It follows easily from the Wang sequence for the projection of M' onto M that $\chi(M) = s \geq 0$. Hence $q(\pi) \geq 0$ also. //

This result could also be proven by counting bases for the cellular chain complex of M' and extending coefficients to $Q(t)$, the field of fractions of $Q\Lambda$.

Theorem 2. *Let M be a closed 4-manifold whose fundamental group π is an extension of Z by a torsion free normal subgroup ν which is isomorphic to the fundamental group of a PD_3-complex N. Then $\pi_2(M) \cong \pi_2(N)$ as $Z[\nu]$-modules if and only if $Hom_{Z[\pi]}(\pi_2(M), Z[\pi]) = 0$. The infinite cyclic covering space M_ν with fundamental group ν is homotopy equivalent to N if and only if $w_1(M)|_\nu = w_1(N)$, $Hom_{Z[\pi]}(\pi_2(M), Z[\pi]) = 0$ and the images of $k_1(M)$ and $k_1(N)$ in $H^3(\nu; \pi_2(M)) \cong H^3(\nu; \pi_2(N))$ generate the same subgroup under* *the action of $Aut_{Z[\nu]}(\pi_2(N))$.*

Proof. If $\Pi = \pi_2(M)$ is isomorphic to $\pi_2(N)$ then it is finitely generated as a $Z[\nu]$-module, by Theorem II.13. As 0 is the only $Z[\pi]$-submodule of $Z[\pi]$ which is finitely generated as a $Z[\nu]$-module it follows that $\Pi^* = Hom_{Z[\pi]}(\pi_2(M), Z[\pi])$ is trivial. It is then clear that the conditions must hold if M_ν is homotopy equivalent to N.

Suppose conversely that these conditions hold. If $\nu = 1$ then M_ν is simply connected and $\pi \cong Z$ has two ends. It follows immediately from Poincaré duality and the UCSS that $H_2(M_\nu; Z) = \Pi \cong \overline{\Pi^*} = 0$ and that $H_3(M_\nu; Z) \cong Z$. Therefore M_ν is homotopy equivalent to S^3. If $\nu \neq 1$ then π has one end, since it has a finitely generated infinite normal subgroup. The hypothesis that $\Pi^* = 0$ implies that $\Pi \cong \overline{H^2(\pi; Z[\pi])}$, by Lemma II.4. Hence $\Pi \cong \overline{H^1(\nu; Z[\nu])}$ as a $Z[\nu]$-module, by the LHSSS. (The overbar notation is unambiguous since $w_1(M)|_\nu = w_1(N)$). But this is isomorphic to $\pi_2(N)$, by Poincaré duality for N. Since N is homotopy equivalent to a 3-dimensional complex the condition on the k-invariants implies that there is a map $f : N \to M_\nu$ which induces isomorphisms on fundamental group and second homotopy group. Since the homology of the universal covering spaces of these spaces vanishes above degree 2 the map f is a homotopy equivalence. //

We do not know whether the hypothesis on the k-invariants is implied by the other hypotheses.

Corollary A. *If moreover N is a 3-manifold whose irreducible factors are Haken, hyperbolic or Seifert fibred then M is homotopy equivalent to a closed PL 4-manifold which fibres over the circle with fibre N.*

Proof. Let $f : N \to M_\nu$ be a homotopy equivalence, where N is a 3-manifold whose irreducible factors are as above. Let $t : M_\nu \to M_\nu$ be the generator of the covering transformations. Then there is a self homotopy equivalence $u : N \to N$ such that $fu \sim tf$. As each irreducible factor of N has the property that self homotopy equivalences are homotopic to PL homeomorphisms [Mo68, Sc83, Wd68], u is homotopic to a homeomorphism [HL74], and so M is homotopy equivalent to the mapping torus of this homeomorphism. //

All known PD_3-complexes are homotopy equivalent to connected sums of Haken, hyperbolic or Seifert fibred 3-manifolds and PD_3-complexes with finite fundamental group, and so in particular have virtually torsion free fundamental group. (However there are PD_3-complexes with finite fundamental group which are not homotopy equivalent to 3-manifolds [Th77]).

Corollary B. *Let M be a closed 4-manifold whose fundamental group π is an extension of Z by a virtually torsion free normal subgroup ν. Then the infinite cyclic covering space M_ν with fundamental group ν is homotopy equivalent to a PD_3-complex if and only if ν is the fundamental group of a PD_3-complex N, $Hom_{Z[\pi]}(\pi_2(M), Z[\pi]) = 0$ and the images of $k_1(M)$ and $k_1(N)$ in $H^3(\nu_o; \pi_2(M)) \cong H^3(\nu_o; \pi_2(N))$ generate the same subgroup under the action of $Aut_{Z[\nu_o]}(\pi_2(N))$, where ν_o is a torsion free subgroup of finite index in ν.*

Proof. The conditions are clearly necessary. Suppose that they hold. Let $\nu_1 \subseteq \nu_o \cap \nu_+ \cap \pi_+$ be a torsion free subgroup of finite index in ν, where $\pi_+ = \mathrm{Ker} w_1(M)$ and $\nu_+ = \mathrm{Ker} w_1(N)$, and let $t \in \pi$ generate π modulo ν. Then each of the conjugates $t^k \nu_1 t^{-k}$ in π has the same index in ν. Since ν is finitely generated the intersection $\mu = \cap t^k \nu_1 t^{-k}$ of all such conjugates has finite index in ν, and is clearly torsion free and normal in the subgroup ρ generated by μ and t. If $\{r_i\}$ is a transversal for ρ in π and $f : \pi_2(M) \to Z[\rho]$ is a nontrivial $Z[\rho]$-linear homomorphism then $g(m) = \Sigma r_i f(r_i^{-1} m)$ defines a nontrivial element of $Hom_\pi(\pi_2(M), Z[\pi])$. Hence $Hom_\rho(\pi_2(M), Z[\rho]) = 0$ and so the covering spaces M_μ and N_ν are homotopy equivalent, by the theorem. It follows easily that M_ν is also a PD_3-complex. //

The condition $Hom_{Z[\pi]}(\pi_2(M), Z[\pi]) = 0$ is often implied by the simpler condition $\chi(M) = 0$. For instance, this is so if ν is free.

Corollary C. *Let M be a closed 4-manifold with $\chi(M) = 0$ and whose fundamental group π is an extension of Z by a normal subgroup ν which is free of finite rank r. Then M is homotopy equivalent to a closed PL 4-manifold which fibres over the circle, with fibre $\sharp^r S^1 \times S^2$ if $w_1(M)|_\nu$ is trivial, and $\sharp^r S^1 \tilde{\times} S^2$ otherwise. The bundle is unique up to bundle isomorphism.*

Proof. Since π is finitely presentable and has cohomological dimension 2 the augmentation module has a resolution $0 \to P_2 \to P_1 \to P_0 \to Z \to 0$ by finitely generated projective left $Z[\pi]$-modules. If we dualize this resolution by means of $P^* = Hom_{Z[\pi]}(P, Z[\pi])$ and use the fact that π has one end we obtain an exact sequence of right $Z[\pi]$-modules $0 \to P_0^* \to P_1^* \to P_2^* \to Ext^2_{Z[\pi]}(Z, Z[\pi]) \to 0$. Since $\chi(\pi) = \chi(M) = 0$ this is a resolution of the right module conjugate to $\pi_2(M)$, by Theorem II.12. On dualizing again we recover the original resolution of Z. Hence $Hom_{Z[\pi]}(\pi_2(M), Z[\pi]) = 0$. Since $c.d.\nu = 1$ and $c.d.\pi = 2$ the first k-invariants of M and N both lie in trivial groups, and so the hypotheses of the theorem hold. The final assertion follows as homotopy implies isotopy for self homeomorphisms of such 3-manifolds [L]. //

The condition $\chi(M) = 0$ also implies $Hom_{Z[\pi]}(\pi_2(M), Z[\pi]) = 0$ if π has one end and $Z[\pi]$ has a safe extension or if π is amenable, by Theorem II.7 and by [Ec93], respectively.

It would also be of interest to replace the condition that ν be isomorphic to the group of a PD_3-complex by more intrinsic, algebraic conditions. In Section 4 we shall give a partial result in this direction.

2. Change of rings and cup products

In the next two sections we shall adapt and extend work of Barge in setting up duality maps in the equivariant (co)homology of covering spaces.

Let π be an extension of Z by a normal subgroup ν and fix an element t of π whose image generates π/ν. Let $\alpha : \nu \to \nu$ be the automorphism determined by $\alpha(h) = tht^{-1}$ for all h in ν. Let $\Lambda = Z[t, t^{-1}]$. The automorphism α extends to a ring automorphism (also denoted by α) of the group ring $Z[\nu]$, and the ring $Z[\pi]$ may then be viewed as a twisted Laurent extension, $Z[\pi] = Z[\nu]_\alpha[t, t^{-1}]$. The ring Λ is the quotient of $Z[\pi]$ by the two-sided ideal generated by $\{h - 1 | h \in \nu\}$, while as a left module over itself $Z[\nu]$ is isomorphic to $Z[\pi]/Z[\pi](t-1)$ and so may be viewed as a left $Z[\pi]$-module. (Note that α is not a module automorphism unless t is central).

If M is a left $Z[\pi]$-module let $M|_\nu$ denote the underlying $Z[\nu]$-module, and let $\hat{M} = Hom_{Z[\nu]}(M|_\nu, Z[\nu])$. Then $\hat{M}$ is a right $Z[\nu]$-module via $(f\xi)(m) =$

$f(m)\xi$ for all ξ in $Z[\nu]$, f in $\hat{M}$ and m in M. If $M = Z[\pi]$ then $\hat{Z}[\pi]$ is also a left $Z[\pi]$-module via $(\phi t^r f)(\xi t^s) = \xi\alpha^{-s}(\phi)f(t^{s-r})$ for all f in $\hat{Z}[\pi]$, ϕ, ξ in ν and r, s in Z. As the left and right actions commute $\hat{Z}[\pi]$ is a $(Z[\pi], Z[\nu])$-bimodule. We may describe this bimodule more explicitly. Let $Z[\nu][[t, t^{-1}]]$ be the set of doubly infinite power series $\Sigma_{n\in Z} t^n \phi_n$ with ϕ_n in $Z[\nu]$ for all n in Z, with the obvious right $Z[\nu]$-module structure, and with the left $Z[\pi]$-module structure given by $\phi t^r(\Sigma t^n \phi_n) = \Sigma t^{n+r}\alpha^{-n-r}(\phi)\phi_n$ for all ϕ, ϕ_n in $Z[\nu]$ and r in Z. (Note that even if $\nu = 1$ this module is not a ring in any natural way). Then the homomorphism $j : \hat{Z}[\pi] \to Z[\nu][[t, t^{-1}]]$ given by $j(f) = \Sigma t^n f(t^n)$ for all f in $\hat{Z}[\pi]$ is a $(Z[\pi], Z[\nu])$-bimodule isomorphism. (Indeed, it is clearly an isomorphism of right $Z[\nu]$-modules, and we have defined the left $Z[\pi]$-module structure on $\hat{Z}[\pi]$ by pulling back the one on $Z[\nu][[t, t^{-1}]]$).

For each f in $\hat{M}$ we may define a function $T_M f : M \to \hat{Z}[\pi]$ by the rule $(T_M f)(m)(t^n) = f(t^{-n}m)$ for all m in M and n in Z. It is routine to check that $T_M f$ is $Z[\pi]$-linear, and that $T_M : \hat{M} \to Hom_{Z[\pi]}(M, \hat{Z}[\pi])$ is an isomorphism of abelian groups. (It is clearly a monomorphism, and if $g : M \to \hat{Z}[\pi]$ is $Z[\pi]$-linear then $g = T_M f$ where $f(m) = g(m)(1)$ for all m in M. In fact if we give $Hom_{Z[\pi]}(M, \hat{Z}[\pi])$ the natural right $Z[\nu]$-module structure by $(\mu\phi)(m) = \mu(m)\phi$ for all ϕ in $Z[\nu]$, $Z[\pi]$-homomorphisms $\mu : M \to \hat{Z}[\pi]$ and m in M them T_M is an isomorphism of right $Z[\nu]$-modules). Thus we have a natural equivalence $T : Hom_{Z[\nu]}(-|_\nu, Z[\nu]) \Rightarrow Hom_{Z[\pi]}(-, \hat{Z}[\pi])$ of functors from $((Mod_{Z[\pi]}))$ to $((Mod_{Z[\nu]}))$. If C_* is a chain complex of left $Z[\pi]$-modules then T induces natural isomorphisms from $H^*(C_*|_\nu; Z[\nu]) = H^*(Hom_{Z[\nu]}(C_*|_\nu, Z[\nu])$ to $H^*(C_*; \hat{Z}[\pi]) = H^*(Hom_{Z[\pi]}(C_*, \hat{Z}[\pi]))$. In particular, since the forgetful functor $-|_\nu$ is exact and takes projectives to projectives there are isomorphisms from $Ext^*_{Z[\nu]}(M|_\nu, Z[\nu])$ to $Ext^*_{Z[\pi]}(M, \hat{Z}[\pi])$ which are functorial in M.

We now define a Z-linear function $e : \Lambda \otimes \hat{Z}[\pi] \to Z[\pi]$ by $e(t^n \otimes f) = t^n f(t^n)$ for all f in $\hat{Z}[\pi]$ and n in Z. If we give $\Lambda \otimes \hat{Z}[\pi]$ the diagonal left $Z[\pi]$-module structure by $\phi t^s(t^r \otimes f) = t^{r+s} \otimes \phi t^s f$ for all f in $\hat{Z}[\pi]$, ϕ in $Z[\nu]$ and r, s in Z then e is $Z[\pi]$-linear. We may use e to define cross products in cohomology. Let A_* be a Λ-chain complex and B_* a $Z[\pi]$-chain complex and give the tensor product the total grading $A_* \otimes B_*$ and differential and the diagonal

$Z[\pi]$-structure. Then there are cross products from $H^p(A_*;\Lambda)\otimes H^q(B_*;\hat{Z}[\pi])$ to $H^{p+q}(A_*\otimes B_*;Z[\pi])$. If we take A_* to be a projective resolution of the Λ-module L and B_* to be a projective resolution of the $Z[\pi]$-module M we obtain cross products from $Ext^p_\Lambda(L,\Lambda)\otimes Ext^q_{Z[\pi]}(M,\hat{Z}[\pi])$ to $Ext^{p+q}_{Z[\pi]}(L\otimes M,Z[\pi])$. In particular, if $L=Z$ we have $Z\otimes M\cong M$ as $Z[\pi]$-modules and $Ext^1_\Lambda(Z,\Lambda)\cong Z$, and so we obtain homomorphisms from $Ext^q_{Z[\pi]}(M,\hat{Z}[\pi])$ to $Ext^{q+1}_{Z[\pi]}(M,Z[\pi])$ to which are functorial in M. Similarly if A_* is the chain complex concentrated in degrees 0 and 1 with $A_0=A_1=\Lambda$ and $\partial_1:A_1\to A_0$ given by multiplication by $t-1$ then there is a chain homotopy equivalence of $A_*\otimes B_*$ with B_* and we obtain homomorphisms from $H^q(B_*;\hat{Z}[\pi])$ to $H^{q+1}(B_*;Z[\pi])$ which are functorial in B_*.

Suppose now that B_* is a free $Z[\pi]$-chain complex such that $B_j=0$ for $j<0$ and $H_0(B_*)\cong Z$. Then the quotient epimorphism from B_0 to $H_0(B_*)$ factors through Λ and there is a chain homomorphism ϵ_* from B_* to the above complex A_*. Let η be the class in $H^1(B_*;\Lambda)=H^1(Hom_{Z[\pi]}(B_*,\Lambda))$ represented by $\epsilon_1:B_1\to\Lambda$. Then η is the image of a generator of the infinite cyclic group $H^1(A_*;\Lambda)=Ext^1_\Lambda(Z,\Lambda)$. Now we may also obtain η as the image of a generator of $H^0(B_*|_\nu;Z)=\{\lambda:B_0\to Z|\lambda \textit{ is } Z[\pi]\textit{-linear and } \lambda\partial_1=0\}\cong Z$ under the analogous pairing from $Ext^1_\Lambda(Z,\Lambda)\otimes H^0(B_*|_\nu;Z)$ to $H^1(B_*;\Lambda)$ of Barge [Ba80']. As $Ext^1_\Lambda(Z,\Lambda)\cong Ext^1_\Lambda(Z,\Lambda)\otimes H^0(B_*|_\nu;Z)\cong Z$, our pairing from $Ext^1_\Lambda(Z,\Lambda)\otimes H^q(B_*;\hat{Z}[\pi])$ to $H^{q+1}(B_*;Z[\pi])$ factors through $H^1(B_*;\Lambda)\otimes H^q(B_*;\hat{Z}[\pi])$ and we may identify our degree raising homomorphisms from $H^q(B_*;\hat{Z}[\pi])$ to $H^{q+1}(B_*;Z[\pi])$ as those given by cup product with η.

3. Duality in infinite cyclic covers

Let E be a PD_4-complex with $\chi(E)=0$ and $f:E\to S^1$ be a map which induces an epimorphism on fundamental groups, with kernel ν. If ν is finite then π has two ends, so $\tilde{M}\simeq S^3$ by Theorem II.10, and $M_\nu=\tilde{M}/\nu$ is a PD_3-complex by [Wl67]. We shall assume henceforth in this section that ν is finitely generated and infinite.

Let C_* be the cellular chain complex of the universal covering space $\tilde{E}$ and fix a generator $[E]$ of $H_4(\bar{Z}\otimes_{Z[\pi]}C_*)\cong Z$. Since π has one end and $\tilde{E}$ is simply connected the only nonzero homology modules are $H_0(C_*) = Z$ and $H_2(C_*) \cong \pi_2(E)$. Since $H_1(\bar{\Lambda} \otimes_{Z[\pi]} C_*) = H_1(E_\nu; Z) \cong \nu/\nu'$ is finitely generated as an abelian group, $Hom_{Z[\pi]}(H_1(\bar{\Lambda} \otimes_{Z[\pi]} C_*), \Lambda) = 0$. An elementary computation then shows that $H^1(C_*; \Lambda)$ is infinite cyclic, and generated by the class η defined in Section 2.

The abelian group $H_3(E_\nu; Z) = H_3(\bar{\Lambda} \otimes_{Z[\pi]} C_*)$ is infinite cyclic, generated by the class $[E_\nu] = \eta \cap [E]$, by Poincaré duality for E with coefficients Λ. Thus $H_q(E_\nu; Z) = H_q(\bar{Z} \otimes_{Z[\nu]} C_*) = H_q(\bar{\Lambda} \otimes_{Z[\pi]} C_*)$ is finitely generated over Z and so is a torsion Λ-module, except perhaps when $q = 2$. On extending coefficients to $Q(t)$, the field of fractions of Λ, we may conclude that $H_2(\bar{\Lambda} \otimes_{Z[\pi]} C_*)$ has rank $\chi(E) = 0$ and so is also a torsion Λ-module. Poincaré duality and the UCSS then imply that $H_2(\bar{\Lambda} \otimes_{Z[\pi]} C_*)$ is isomorphic to $\overline{Ext^1_\Lambda(H_1(\bar{\Lambda} \otimes_{Z[\pi]} C_*), \Lambda)}$, and so is finitely generated and torsion free over Z. Therefore E_ν satisfies Poincaré duality with coefficients Z, i.e., cap product with $[E_\nu]$ maps $\overline{H^p(C_*; Z)}$ isomorphically to $H_{3-p}(\bar{Z} \otimes_{Z[\nu]} C_*) = H_{3-p}(\bar{\Lambda} \otimes_{Z[\pi]} C_*)$.

The complex $C_*|_\nu$ is the cellular chain complex for $\tilde{E}$, considered as the universal covering space of E_ν. Therefore, by standard properties of cap and cup products, to show that E_ν satisfies Poincaré duality with local coefficients $Z[\nu]$, i.e., that cap product with $[E_\nu]$ gives isomorphisms from $\overline{H^p(C_*; Z[\nu])}$ to $H_{3-p}(C_*)$, it shall suffice to show that the map η_p from $\overline{H^p(C_*; Z[\nu])} = \overline{H^p(C_*; \hat{Z}[\pi])}$ to $\overline{H^{p+1}(C_*; Z[\pi])}$ given by cup product with η is an isomorphism, for all p. There is nothing to prove when $p = 0$ or $p > 4$, for both modules are then 0.

The cohomology modules $H^p(C_*; Z[\nu])$ and $H^p(C_*; Z[\pi])$ may be "computed" via the UCSS. In this case the E_2^{p*} columns are nonzero only for $p = 0$ or 2. In particular, $H^1(C_*; Z[\nu]) \cong Ext^1_{Z[\nu]}(Z, Z[\nu])$, and, if $Hom_{Z[\pi]}(\Pi, Z[\pi])$ is 0, then $H^2(C_*; Z[\pi]) \cong Ext^2_{Z[\pi]}(Z, Z[\pi])$. Cup product with η determines homomorphisms between these spectral sequences which are compatible with the corresponding homomorphisms between the cohomology modules. The E_2^{0*} terms of these spectral sequences involve only the cohomology of the groups and the homomorphisms between them may be identified with the maps arising in the LHSSS for π as an extension of Z by ν.

In particular, as ν is finitely generated $H^1(\nu; Z[\pi]) \cong \Lambda \otimes H^1(\nu; Z[\nu])$ and so $H^1(\pi/\nu; H^1(\nu; Z[\pi])) = H^1(\nu; Z[\nu]) = Ext^1_{Z[\nu]}(Z, Z[\nu])$. If moreover ν is almost finitely presentable then $H^2(\nu; Z[\pi]) \cong \Lambda \otimes H^2(\nu; Z[\nu])$ and so $H^0(\pi/\nu; H^2(\nu; Z[\pi])) = 0$. It then follows that the homomorphism from $Ext^1_{Z[\nu]}(Z, Z[\nu])$ to $Ext^2_{Z[\pi]}(Z, Z[\pi])$ is an isomorphism, and so η_1 is an isomorphism if $Hom_{Z[\pi]}(\Pi, Z[\pi]) = 0$.

4. Virtual products

In this section we shall show directly that, under suitable hypotheses on the fundamental group, an infinite cyclic covering of a closed 4-manifold with Euler characteristic 0 satisfies Poincaré duality with local coefficients, by modifying the arguments used by Barge in establishing duality with constant coefficients [Ba80,80']. The following strategy may also be used to give a different proof of Corollary C of Theorem 2 above (see [Hi89]).

Theorem 3. *Let M be a closed 4-manifold with $\chi(M) = 0$ and whose fundamental group π is an extension of Z by an infinite normal subgroup ν. If π has an infinite cyclic normal subgroup C which is not contained in ν then the covering space M_ν with fundamental group ν is a PD_3-complex.*

Proof. We may assume without loss of generality that M is orientable and that C is central in π. Since $C \cap \nu = 1$ the subgroup $C\nu \cong C \times \nu$ has finite index in π. Thus by passing to a finite cover we may assume that $\pi = C \times \nu$, and hence that $Z[\pi] = Z[\nu][t, t^{-1}]$. It follows immediately that ν is finitely presentable and that π has one end. Since M_ν is an open 4-manifold it is smoothable, by Theorem VIII.2 of [FQ], and so is homotopy equivalent to a 3-dimensional complex. Therefore the chain complex C_* of M_ν is chain homotopy equivalent over $Z[\nu]$ to a complex which is concentrated in degrees 0, 1, 2 and 3. In particular, $H^4(C_*; Z[\nu]) = 0$. Let $\Pi = \pi_2(M)$. The localization of $Z[\pi]$ with respect to the central multiplicative system $S = \{(t-1)^n | n \geq 1\}$ is a safe extension. Therefore $Hom_{Z[\pi]}(\Pi, Z[\pi]) = 0$, by Theorem II.7, and so η_1 is an isomorphism, by the final remark of the previous section. It remains to be shown that $H^2(C_*; Z[\nu]) = 0$ and that η_3 is an isomorphism.

Let $Lexp_z(r(t))$ denote the Laurent expansion of the rational function $r(t)$ at z and let $T : Z[\pi]_S \to \hat{Z}[\pi]$ be the function defined by the equation $T(\gamma/(t-1)^n) = \gamma(Lexp_\infty((t-1)^{-n}) - Lexp_0((t-1)^{-n}))$, for all γ in $Z[\pi]$ and $n \geq 0$. Then $T(1/(t-1)^{k+1}) = \Sigma_{m\geq 0} C(m+k,k)(t^{-m-k-1} + (-1)^k t^m)$ where $C(m+k,k) = (m+k)!/m!k!$ is the binomial coefficient. (In particular, $T(1/(t-1)) = \Sigma_{n\in Z} t^n$). It follows easily that $(t-1)T(1/(t-1)^{k+1}) = T(1/(t-1)^k)$ and hence that T is well defined, $Z[\pi]$-linear, the kernel of T contains $Z[\pi]$ and the image of T is contained in the S-torsion submodule of $\hat{Z}[\pi]$.

Suppose that $T(\gamma/(t-1)^n) = 0$. Then $T(\gamma/(t-1)^j) = 0$ for all $1 \leq j \leq n$. Hence on writing $\gamma = \Sigma_{i\geq 0} a_i (t-1)^i$ a simple induction gives $a_i = 0$ for $0 \leq i < n$. Thus $\mathrm{ker}T = Z[\pi]$. Suppose next that $(t-1)\Sigma_{n\in Z} a_n t^n = 0$. Then $a_n = a_{n+1}$ for all n in Z, and so $\Sigma_{n\in Z} a_n t^n = T(a_0/(t-1))$. It follows easily that $T(Z[\pi]_S)$ is the S-torsion $Z[\pi]$-submodule of $\hat{Z}[\pi]$, and so $\Omega = \mathrm{coker}T$ is S-torsion free. It is easily verified that $(t-1)\hat{Z}[\pi] = \hat{Z}[\pi]$. Hence $t-1$ acts invertibly on Ω, and so $\Omega_S = \Omega$. Thus there is a 4-term exact sequence of $Z[\pi]$-modules

$$0 \to Z[\pi] \to Z[\pi]_S \xrightarrow{T} \hat{Z}[\pi] \to \Omega \to 0$$

in which two of the terms are S-divisible. (Note that neither $Z[\pi]$ nor $\hat{Z}[\pi]$ are $Z[\pi]_S$-modules).

Since $Z_S = \Pi_S = 0$, by Theorem 7, the localized chain complex C_{*S} is contractible and so $H^*(C_*; Z[\pi]_S) = H^*(C_{*S}; Z[\pi]_S) = 0$ and $H^*(C_*; \Omega) = H^*(C_{*S}; \Omega) = 0$. Hence $H^n(C_*; \hat{Z}[\pi])$ and $H^{n+1}(C_*; Z[\pi])$ are isomorphic for all n. In particular, $H^2(C_*; Z[\nu]) = H^2(C_*; \hat{Z}[\pi]) = H^3(C_*; Z[\pi]) = 0$ and $H^3(C_*; Z[\nu]) \cong H^4(C_*; Z[\pi]) \cong Z$. As C_* is chain homotopy equivalent over $Z[\nu]$ to a 3-dimensional complex, the map from $H^3(C_*; Z[\nu])$ to $H^3(C_*; Z)$ induced by the augmentation of $Z[\nu]$ onto Z is surjective.

Therefore cap product with $[M_\nu]$ maps $H^3(C_*; Z)$ onto $H_0(\bar{Z}\otimes_{Z[\nu]} C_*) \cong Z$. Since the composed map from $H^3(C_*; Z[\nu]) \cong Z$ to $Z \cong H_0(C_*)$ may also be viewed as cap product with M_ν, we conclude that the latter map is an isomorphism. Thus the space M_ν satisfies Poincaré duality with coefficients $Z[\nu]$. Since ν is finitely presentable, M_ν is then finitely dominated [Br72], and so is a PD_3-complex. //

Corollary. *Let M be a closed 4-manifold with $\chi(M) = 0$ and whose fundamental group π is an extension of Z by an almost finitely presentable normal subgroup ν. If ν is finite then it has cohomological period dividing 4. If ν has one end and either $Z[\pi]$ has a safe extension or π is amenable then M is aspherical and so π is a PD_4-group. If ν has two ends then $\nu \cong Z$, $Z \oplus (Z/2Z)$ or $D = (Z/2Z) * (Z/2Z)$.*

Proof. The first assertion was proven in Theorem II.10. If ν has one end then an LHSSS argument shows that $H^s(\pi; Z[\pi]) = 0$ for $s \leq 2$ and so the second assertion follows from the Corollary to Theorem II.7, if $Z[\pi]$ has a safe extension, and from [Ec93], if π is amenable. If ν has two ends then it is an extension of Z or $(Z/2Z) * (Z/2Z)$ by a finite normal subgroup, and so $Out(\nu)$ is finite. Therefore the centre of π is not contained in ν, so M_ν is a PD_3-complex, and the final assertion follows from Theorem 4.4 of [Wl67]. //

In Chapter V we shall strengthen this Corollary to obtain a fibration theorem for 4-manifolds with torsion free elementary amenable fundamental group.

The theorem remains true if C is contained in ν.

Theorem 4. *Let M be a closed 4-manifold with $\chi(M) = 0$ and whose fundamental group π is an extension of Z by an almost finitely presentable normal subgroup ν. If ν has an infinite cyclic subgroup C which is normal in π then the covering space M_ν with fundamental group ν is a PD_3-complex. In particular, if $[\nu : C] = \infty$ then M is aspherical and ν is a PD_3-group.*

Proof. Since a cell complex is a PD_n-complex if and only if some finite covering space is so, we may assume without loss of generality that M is orientable and that C is central in π. If C has finite index in ν then we may also assume that $\nu = C$. But then the centre of π is not contained in ν and so the result follows from Theorem 3.

We may thus also assume henceforth that $[\nu : C] = \infty$. An LHSSS argument shows that $H^2(\pi; Z[\pi]) = 0$ for $s \leq 2$, and so M is aspherical and π is a PD_4-group, by the Corollary to Theorem II.7. Since the index $[\pi : \nu]$ is infinite $c.d.\nu \leq 3$ [St77] and so the augmentation $Z[\nu]$-module Z has a projective resolution

$$0 \to P_3 \to P_2 \to P_1 \to P_0 \to Z \to 0.$$

Since ν is FP_2 we may assume that the modules P_i are finitely generated for $i \leq 2$. Moreover the natural homomorphisms from $H^i(\nu; Z[\nu]) \otimes \Lambda$ to $H^i(\nu; Z[\pi])$ are isomorphisms for $i \leq 2$.

The LHSSS for ν as an extension of $G = \nu/C$ by C gives an isomorphism $H^3(\nu; Z[\nu]) \cong H^2(G; H^1(C; Z[\nu])) \cong H^2(G; Z[G])$, since C is finitely generated. Likewise we have $H^3(\nu; Z[\pi]) \cong H^2(G; Z[G]^{(Z)}) \cong H^2(G; Z[G])^{(Z)}$ since $Z[\pi] \cong Z[\nu]^{(Z)}$ and G is FP_2. Therefore $H^3(\nu; Z[\pi]) \cong H^3(\nu; Z[\nu])^{(Z)}$. On keeping track of the direct sum decompositions we may conclude that in fact $H^3(\nu; Z[\pi]) \cong H^3(\nu; Z[\nu]) \otimes \Lambda$ as a Λ-module.

The LHSSS for π as an extension of Z by ν, with coefficients $Z[\pi]$, reduces to a Wang sequence

$$\ldots \to H^q(\nu; Z[\pi]) \to H^q(\nu; Z[\pi]) \to H^{q+1}(\pi; Z[\pi]) \to \ldots$$

Using the above information on $H^q(\nu; Z[\pi])$ we find that $H^q(\nu; Z[\nu]) = 0$ for $q \neq 3$ and $H^3(\nu; Z[\nu]) \cong H^4(\pi; Z[\pi]) \cong Z$. Thus if we dualize the above $Z[\nu]$-resolution of Z by means of $P^* = Hom_{Z[\nu]}(P, Z[\nu])$ we get an exact sequence

$$0 \to P_0^* \to P_1^* \to P_2^* \to P_3^* \to H^3(\nu; Z[\nu]) \cong Z \to 0.$$

The dual of a projective module P is finitely generated if and only if P is Therefore P_3^* and hence P_3 are finitely generated, and so ν is FP_3. As $H^q(\nu; Z[\nu]) \cong Z$ if $q = 3$ and is 0 otherwise ν is a PD_3-group. //

If such a subgroup ν has a subgroup of finite index with infinite abelianization then it is in fact the fundamental group of an aspherical closed Seifert fibred 3-manifold [Hi85]. (See also the Appendix).

We may now show that the conditions on the fundamental group and Euler characteristic which are obviously necessary are almost sufficient alone to characterize products up to homotopy equivalence.

Theorem 5. *Let M be a closed 4-manifold whose fundamental group π is torsion free. Then M is homotopy equivalent to the cartesian product of a closed 3-manifold and S^1 if and only if $\chi(M) = 0$, there is an isomorphism*

$\theta : \pi \to \nu \times Z$ such that $w_1(M)\theta^{-1}|_Z = 0$, where ν is a torsion free 3-manifold group, and $w_2(M) = w_1(M)^2$.

Proof. The conditions are clearly necessary. Suppose that they hold. The covering space M' with fundamental group ν is a PD_3-complex, by Theorem 3 above. The indecomposable free factors of ν are either the groups of aspherical closed 3-manifolds or infinite cyclic, and so by [Tu90] there is a homotopy equivalence $h : M' \to N$, where N is a closed 3-manifold. Let ϕ be a generator of the covering group $Aut(M/M') \cong Z$. The manifold M may be recovered up to homotopy equivalence as the mapping torus $M(\psi)$, where ψ is the self homotopy equivalence of N defined by $\psi = h\phi h^{-1}$. We may assume that ψ fixes a basepoint and that it induces the identity on $\pi_1(N)$, since $\pi_1(M) \cong \nu \times Z$. Moreover ψ preserves the local orientation, since $w_1(M)\theta^{-1}|_Z = 0$. Since ν has no element of order 2 N has no two-sided projective planes and so ψ is homotopic to a rotation about a 2-sphere [He]. Since $w_2(M) = w_1(M)^2$ the rotation is homotopic to the identity and so M is homotopy equivalent to $N \times S^1$. //

Let τ be the twist map of $S^1 \times S^2$, given by $\tau(x, y) = (x, \rho(x)(y))$ for all (x, y) in $S^1 \times S^2$, where ρ is an essential map from S^1 to $SO(3)$. The mapping torus $M(\tau)$ has fundamental group $Z \times Z$ and Euler characteristic 0 but $w_2(M(\tau)) \neq 0$ and $M(\tau)$ is not homotopy equivalent to a product. (Clearly however $M(\tau^2) = S^1 \times S^2 \times S^1$).

5. Circle bundles

In this section we shall give necessary conditions for a closed 4-manifold M to be homotopy equivalent to the total space of an S^1-bundle with base a closed 3-manifold N, and shall show that they are also sufficient if $\pi_1(N)$ is torsion free but not free and if the connecting homomorphism from $\pi_2(N)$ to $\pi_1(S^1)$ is trivial.

In general, any S^1-bundle over a connected base B is induced from some bundle over $P_2(B)$. Given an epimorphism $\gamma : \mu \to \nu$ with cyclic kernel and such that the action of ν on Kerγ determined by conjugation in μ factors through multiplication by ± 1 there is an S^1-bundle $p(\gamma) : X(\gamma) \to Y(\gamma)$ whose fundamental group sequence realizes γ and which is universal for such bundles; the total space of this universal bundle is a $K(\mu, 1)$ space (cf. Proposition 11.4 of [W]).

The conditions in the next lemma are not all independent.

Lemma 6. *Let $p : E \to B$ be the projection of an S^1-bundle ξ over a connected finite complex B. Then*
(i) $\chi(E) = 0$;
(ii) the natural map $p_* : \pi = \pi_1(E) \to \nu = \pi_1(B)$ is an epimorphism with cyclic kernel, and the action of ν on $\mathrm{Ker}p_*$ induced by conjugation in π is given by $w = w_1(\xi) : \pi_1(B) \to Z/2Z \cong \{\pm 1\} \leq Aut(\mathrm{Ker}p_*)$;
(iii) if B is a PD-complex $w_1(E) = p^*(w_1(B) + w)$;
(iv) if B is a PD_3-complex there are maps $y : P_2(B) \to Y(p_*)$ and $\hat{c} : E \to P_2(B)$ such that $c_{P_2(B)} = c_{Y(p_*)}y$, $y\hat{c} = p(p_*)c_E$ and $(\hat{c}, c_E)_*[E] = \pm G(f_{B*}[B])$ where G is the Gysin homomorphism from $H_3(P_2(B); Z^{w_1(B)})$ to $H_4(P_2(E); Z^{w_1(E)})$;
(v) If B is a PD_3-complex $c_{E*}[E] = \pm G(c_{B*}[B])$, where G is the Gysin homomorphism from $H_3(\nu; Z^{w_B}) = H_3(\nu; H_1(\mathrm{Ker}\gamma; Z^{w_E}))$ to $H_4(\pi; Z^{w_E})$;
(vi) $\mathrm{Ker}p_*$ acts trivially on $\pi_2(E)$.

Proof. Condition(i) follows from the multiplicativity of the Euler characteristic in a fibration. If α is any loop in B the total space of the induced bundle $\alpha^*\xi$ is the torus if $w(\alpha) = 0$ and the Klein bottle if $w(\alpha) = 1$ in $Z/2Z$; hence $gzg^{-1} = z^{\epsilon(g)}$ where $\epsilon(g) = (-1)^{w(p_*(g))}$ for g in $\pi_1(E)$ and z in $\mathrm{Ker}p_*$. Conditions (ii) and (vi) then follow from the exact homotopy sequence. If the base B is a PD-complex then so is E, and we may use naturality and the Whitney sum formula (applied to the Spivak normal bundles) to show that $w_1(E) = p^*(w_1(B) + w_1(\xi))$. (As $p^* : H^1(B; Z/2Z) \to H^1(E; Z/2Z)$ is a monomorphism this equation determines $w_1(\xi)$).

Condition (iv) implies (v), and follows from the observations in the paragraph preceding the lemma. (Note that the Gysin homomorphisms G in (iv) and (v) are well defined, since $H_1(\mathrm{Ker}\gamma; Z^{w_E})$ is isomorphic to Z^{w_B}, by (iii)). //

Bundles for which $\mathrm{Ker}p_* \cong Z$ have the following equivalent characterizations.

Lemma 7. *Let ξ be an S^1-bundle with total space E, base B and projection $p: E \to B$, and suppose that B is connected. Then the following conditions are equivalent:*
(i) ξ is induced from an S^1-bundle over $K(\pi_1(B),1)$ via c_B;
(ii) for each map $\beta: S^2 \to B$ the induced bundle $\beta^*\xi$ is trivial;
(iii) the kernel of the epimorphism $p_*: \pi_1(E) \to \pi_1(B)$ induced by p is infinite cyclic.
If these conditions hold then $c(\xi) = c_B^*\Xi$, where $c(\xi)$ is the characteristic class of ξ in $H^2(B; Z^w)$ and Ξ is the class of the extension of fundamental groups in $H^2(\pi_1(B); Z^w) = H^2(K(\pi_1(B),1); Z^w)$, where $w = w_1(\xi)$.

Proof. Condition (*i*) implies condition (*ii*) as for any such map β the composite $c_B\beta$ is nullhomotopic. Conversely, as we may construct $K(\pi_1(B),1)$ by adjoining cells of dimension ≥ 3 to B condition (*ii*) implies that we may extend ξ over the 3-cells, and as S^1-bundles over S^n are trivial for all $n > 2$ we may then extend ξ over the whole of $K(\pi_1(B),1)$, so that (*ii*) implies (*i*). The equivalence of (*ii*) and (*iii*) follows on observing that (*iii*) holds if and only if $\partial\beta = 0$ for all such β, where ∂ is the connecting map from $\pi_2(B)$ to $\pi_1(S^1)$ in the exact sequence of homotopy for ξ, and on comparing this with the corresponding sequence for $\beta^*\xi$.

As the natural map from the set of S^1-bundles over $K(\pi,1)$ with $w_1 = w$ (which are classified by $H^2(K(\pi,1); Z^w)$) to the set of extensions of π by Z with π acting via w (which are classified by $H^2(\pi; Z^w)$) which sends a bundle to the extension of fundamental groups is an isomorphism we have $c(\xi) = c_B^*(\Xi)$. //

If N is a closed 3-manifold which has no summands of type $S^1 \times S^2$ or $S^1 \tilde{\times} S^2$ (i.e., if $\pi_1(N)$ has no infinite cyclic free factor) then every S^1-bundle over N with $w = 0$ restricts to a trivial bundle over any map from S^2 to N. For if ξ is such a bundle, with characteristic class $c(\chi)$ in $H^2(N; Z)$, and $\beta: S^2 \to N$ is any map then $\beta_*(c(\beta^*\xi) \cap [S^2]) = \beta_*(\beta^*c(\xi) \cap [S^2]) = c(\xi) \cap \beta_*[S^2] = 0$, as the Hurewicz homomorphism is trivial for such N. Since

β_* is an isomorphism in degree 0 it follows that $c(\beta^*\xi) = 0$ and so $\beta^*\xi$ is trivial. (A similar argument applies for bundles with $w \neq 0$, provided the induced 2-fold covering space N^w has no summands of type $S^1 \times S^2$ or $S^1 \tilde{\times} S^2$).

On the other hand the bundle with total space $S^1 \times S^3$, base $S^1 \times S^2$ and projection $id_{S^1} \times \eta$ (where η is the Hopf fibration) is clearly nontrivial when pulled back over any esssential map from S^2 to $S^1 \times S^2$, and is not induced from any bundle over $K(Z,1)$. Moreover, a closed 3-manifold N with no summands of type $S^1 \times S^2$ or $S^1 \tilde{\times} S^2$ may have a 2-fold covering space (corresponding to a homomorphism w from $\pi_1(N)$ to $Z/2Z$) with such summands; for instance $S^1 \times S^2$ is a 2-fold covering space of $RP^3 \sharp RP^3$.

Theorem 8. *Let M be a finite PD_4-complex and N a finite PD_3-complex whose fundamental group is torsion free but not free. Then M is homotopy equivalent to the total space of an S^1-bundle over N which satisfies the conditions of Lemma 7 if and only if (i) $\chi(M) = 0$;*

(ii) there is an epimorphism γ from $\pi = \pi_1(M)$ to $\nu = \pi_1(N)$ with infinite cyclic kernel;

(iii) $w_1(M) = (w_1(N)+w)\gamma$, where $w : \nu \to Z/2Z \cong Aut(\mathrm{Ker}\gamma)$ is determined by the action of ν on $\mathrm{Ker}\gamma$ induced by conjugation in π;

(iv) there is a map $\hat{c} : M \to P_2(N)$ such that $c_{P_2(N)}\hat{c} = p(\gamma)c_M$ and $(\hat{c}, c_M)_*[M] = \pm G(f_{N*}[N])$ in $H_4(P_2(N) \times_{K(\nu,1)} K(\pi,1); Z^{w_1(M)})$, where G is the Gysin homomorphism.

Moreover, if these conditions hold then M has minimal Euler characteristic for its fundamental group, i.e. $q(\pi) = 0$.

Proof. Since these conditions are homotopy invariant and hold if M is the total space of such a bundle, they are necessary. Suppose conversely that they hold. As ν is torsion free N is the connected sum of a 3-manifold with free fundamental group and some aspherical PD_3-complexes [Tu90]. As ν is not free there is at least one aspherical summand. Hence $c.d.\nu = 3$ and $H_3(c_N; Z^{w_1(N)})$ is a monomorphism.

Let $p(\gamma) : K(\pi,1) \to K(\nu,1)$ be the S^1-bundle corresponding to γ and let $E = N \times_{K(\nu,1)} K(\pi,1)$ be the total space of the S^1-bundle over N induced by the classifying map $c_N : N \to K(\nu,1)$. The bundle map covering c_N is the classifying map c_E. Then $\pi_1(E) \cong \pi = \pi_1(M)$, $w_1(E) = (w_1(N)+w)\gamma = w_1(M)$, as maps from π to $Z/2Z$, and $\chi(E) = 0 = \chi(M)$, by conditions (i) and (iii). The maps c_N and c_E induce a homomorphism between the Gysin sequences of the S^1-bundles. Since N and ν have homological dimension 3 it follows easily that $H_4(c_E; Z^{w_1(E)})$ is a monomorphism, and hence *a fortiori* that $H_4(f_E; Z^{w_1(E)})$ is also a monomorphism. Since condition (iv) gives us a map $(\hat{c}, c_M)$ from M to $P_2(N) \times_{K(\nu,1)} K(\pi,1) \simeq P_2(E)$ such that $(\hat{c}, c_M)_*[M] = \pm f_{E*}[E]$ it follows from Theorem II.8 that M is homotopy equivalent to E.

If $\pi_1(M)$ has an infinite cyclic normal subgroup of infinite index then π has one end and $Z[\pi]$ has a safe extension, by Theorem I.7, so $\chi(M) \geq 0$, by Theorem II.7. This implies the final assertion. //

Note that the Gysin homomorphism is here an isomorphism, since $c.d.\nu = 3$. We suspect that there may be a better formulation of this theorem, in which condition (iv) is replaced by other, more transparent conditions, and we shall give some evidence for this expectation in the next paragraphs.

As π is an extension of ν by $\mathrm{Ker}\gamma \cong Z$ and $c.d.\nu = 3$ we have $c.d.\pi = 4$, by Theorem 5.6 of [Bi]. As π has one end and $Z[\pi]$ has a safe extension $\pi_2(M) \cong \overline{H^2(\pi; Z[\pi])}$, by Theorem II.7. It follows from conditions (ii) - (iv) and the LHSSS that $\pi_2(M) \cong \pi_2(E) \cong \gamma^*\pi_2(N) \otimes_Z Z^w$ as $Z[\pi]$-modules, where $\gamma^*\pi_2(N)$ denotes the $Z[\pi]$-module with the same underlying group and with $Z[\pi]$-action determined by the homomorphism γ. If the first k-invariants of M and E agree the spaces $P_2(M)$ and $P_2(E)$ are homotopy equivalent.

As $\pi_2(N)$ is a projective $Z[\nu]$-module, by Theorem II.13, $H^q(\pi; \gamma^*\pi_2(N) \otimes_Z Z^{w_1(N)}) = 0$ if $q \geq 2$. Hence it follows from the spectral sequence for $c_{P_2(M)}$ that $H_4(P_2(M); Z^{w_1(M)})$ maps onto $H_4(\pi; Z^{w_1(M)})$, with kernel isomorphic to $H_0(\pi; \Gamma(\pi_2(M))) \otimes_Z Z^{w_1(M)})$, where $\Gamma(\pi_2(M)) = H_4(K(\pi_2(M), 2); Z)$ is Whitehead's universal quadratic construction on $\pi_2(M)$. This suggests that under the other assumptions of the theorem condition (iv) may be equivalent to conditions (v) and (vi) of Lemma 6 together with some information on the first k-invariant and the intersection pairing on $\pi_2(M)$.

If $\nu = 1$ then M is orientable, $\pi \cong Z$ and $\chi(M) = 0$, so $M \simeq S^3 \times S^1$. If

ν is a nontrivial free group then $c.d.\pi = 2$ and (v) is vacuous. In general, if the restriction on ν is removed it is not clear that there should be a degree 1 map from M to the total space of such a bundle.

It would be of interest to have a theorem that involved only assumptions on M, without specifying N in advance. For instance, the conditions "$\chi(M) = 0$" and "$\pi_1(M)$ has an infinite cyclic normal subgroup A such that $\pi_1(M)/A$ is virtually of finite cohomological dimension" are certainly necessary for M to be homotopy equivalent to the total space of an S^1-bundle over some 3-manifold. Are they sufficient? In the aspherical case there is such a characterization, modulo the question of whether every PD_3-group is a 3-manifold group.

Theorem 9. *A finite PD_4-complex M is homotopy equivalent to the total space of an S^1-bundle over an aspherical PD_3-complex if and only if $\chi(M) = 0$ and $\pi = \pi_1(M)$ has an infinite cyclic normal subgroup A such that π/A has one end and finite cohomological dimension.*

Proof. The conditions are clearly necessary. Conversely, suppose that they hold. Since π/A has one end $H^s(\pi/A; Z[\pi/A]) = 0$ for $s \leq 1$ and so an LHSSS calculation gives $H^t(\pi; Z[\pi]) = 0$ for $t \leq 2$. Moreover $Z[\pi]$ has a safe extension, by Theorem I.7. Hence M is aspherical and π is a PD_4-group, by the Corollary to Theorem II.7. Since A is FP_∞ and $c.d.\pi/A < \infty$ the quotient π/A is a PD_3-group, by Theorem 9.11 of [Bi]. Therefore M is homotopy equivalent to the total space of an S^1-bundle over the PD_3-complex $K(\pi/A, 1)$. //

Note that a finitely generated torsion free group has one end if and only if it is indecomposable as a free product and is neither infinite cyclic nor trivial. In particular, M is homotopy equivalent to a product of an aspherical PD_3-complex with S^1 if and only if $\chi(M) = 0$ and $\pi_1(M) \cong \nu \times Z$ where ν has one end and finite cohomological dimension.

CHAPTER IV

SURFACE BUNDLES

If B, E and F are connected finite complexes and $p : E \to B$ is a Hurewicz fibration with fibre homotopy equivalent to F then $\chi(E) = \chi(B)\chi(F)$ and there is an exact sequence $\pi_2(B) \to \pi_1(F) \to \pi_1(E) \to \pi_1(B) \to 1$ in which the image of $\pi_2(B)$ is in the centre of $\pi_1(F)$. These conditions are clearly homotopy invariant.

In this chapter we shall show that a closed 4-manifold M is homotopy equivalent to the total space of a fibre bundle with base and fibre closed surfaces if and only if these conditions on the Euler characteristic and fundamental group hold. When the base is S^2 or RP^2 we need also conditions on the characteristic classes of M, and in the latter case our results are incomplete when M is nonorientable.

1. Bundles with base an aspherical surface

If X is a finite complex let $E(X)$ be the monoid of all self homotopy equivalences of X, with the compact-open topology, and if X has been given a base point let $E_0(X)$ be the subspace (and submonoid) of base point preserving self homotopy equivalences. The evaluation map from $E(X)$ to X is a Hurewicz fibration with fibre $E_0(X)$ [Go68]. If B and X are finite complexes then Hurewicz fibrations with base B and fibre X are classified by homotopy classes of maps from B to the Milgram classifying space $BE(X)$ of $E(X)$ [Mi67].

Theorem 1. *Let M be a closed 4-manifold and B and F aspherical closed surfaces. Then M is homotopy equivalent to the total space of an F-bundle over B if and only if $\chi(M) = \chi(B)\chi(F)$ and $\pi_1(M)$ is an extension of $\pi_1(B)$*

Typeset by $\mathcal{A}\mathcal{M}\mathcal{S}$-TEX

by $\pi_1(F)$. Moreover every extension of $\pi_1(B)$ by $\pi_1(F)$ is realized by some surface bundle, which is determined up to isomorphism by the extension.

Proof. The conditions are clearly necessary. Suppose that they hold. Then there is an epimorphism θ from $\pi_1(M)$ to $\pi_1(B)$ with kernel isomorphic to $\pi_1(F)$. Since B is aspherical there is a map h from $K(\pi_1(M), 1)$ to B which induces θ. The homotopy fibre of h is clearly homotopy equivalent to F. If $\chi(F) < 0$ then the identity components of $Diff(F)$ and $E(F)$ are contractible. The torus and the Klein bottle admit natural actions of $S^1 \times S^1$ and S^1, respectively, and the induced maps from these groups to the identity components of $Diff(F)$ and $E(F)$ are also homotopy equivalences. (See page 21 of [EE69] and Theorem III.2 of [Go65]). Moreover $\pi_0(E(F)) \cong Out(\pi_1(F))$, in all cases. Now every automorphism of $\pi_1(F)$ is realizable by a diffeomorphism and homotopy implies isotopy for self diffeomorphisms of surfaces. (See Chapter V of [ZVC]). Hence the inclusion of $Diff(F)$ into $E(F)$ is a homotopy equivalence. Therefore h is fibre homotopy equivalent to the projection of a fibre bundle, which is unique up to bundle isomorphism. The total space E of such a bundle is a $K(\pi_1(M), 1)$ space, and $\chi(E) = \chi(B)\chi(F) \geq 0$. Therefore $H^2(E; Z/2Z) \neq 0$ and so it follows from Theorem II.6 that the classifying map $c : M \to E$ is a homotopy equivalence.

If $\chi(F) < 0$ then $BDiff(F) \simeq BE(F) \simeq K(Out(\pi_1(F), 1)$, so F-bundles over B are classified by $[B, K(Out(\pi_1(F), 1))] = Hom(\pi_1(B), Out(\pi_1(F)))$. Since $\zeta\pi_1(F) = 1$ each such homomorphism corresponds to an unique equivalence class of extensions of $\pi_1(B)$ by $\pi_1(F)$, by Proposition 11.4.21 of [Ro].

If $F = S^1 \times S^1$ then F acts on itself by left multiplication. Let $\alpha : GL(2, Z) \to Aut(F) \leq Diff(F)$ be the standard linear action. Then the natural maps from the semidirect product $F \times_\alpha GL(2, Z)$ to $Diff(F)$ and to $E(F)$ are homotopy equivalences. Therefore $BDiff(F)$ is a $K(Z^2, 2)$-fibration over $K(GL(2, Z), 1)$. It follows that F-bundles over B are classified by two invariants: a homomorphism $\xi : \pi_1(B) \to GL(2, Z)$ together with a cohomology class in $H^2(B; (Z^2)^\xi)$, where $(Z^2)^\xi$ is the $Z[\pi_1(B)]$-module determined by ξ. Every such action is realizable by an extension (for instance, the semidirect product $Z^2 \times_\xi \pi_1(B)$) and the extensions realizing ξ are classified up to equivalence by $H^2(\pi_1(B); (Z^2)^\xi)$. Thus if B is aspherical we again see that the natural map from bundles to group extensions is a bijection. As a similar argument applies if F is the Klein bottle this proves the theorem. //

If $\chi(B) < \chi(F)$ then the the image of $\pi_1(F)$ is a characteristic subgroup of $\pi_1(M)$. (This is clear if $\chi(F) < 0$, and is proven in [Jo79] otherwise). Hence if the total space of two such bundles ξ and ξ' have isomorphic fundamental groups then $\xi = f^*\xi'$ for some diffeomorphism $f : B \to B$. In particular, the total spaces are diffeomorphic. (Compare the statement of Melvin's theorem given later in this section).

In the following lemma we shall use the more usual notation G_3 for $E(S^2)$ and SG_3 for the submonoid of degree 1 maps, i.e., the connected component of id_{S^2} in G_3. The connected component of id_{S^2} in $E_0(S^2)$ may be identified with the double loop space $\Omega^2 S^2$.

Lemma 2. *Let X be a finite 2-complex. Then the maps from $[X; BO(3)]$ to $[X; BG_3]$ and from $[X; BSO(3)]$ to $[X; BE(RP^2)]$ induced by the inclusions of $O(3)$ into G_3 and $SO(3)$ into $E(RP^2)$ are surjective.*

Proof. It shall suffice to prove that the maps from $O(3)$ to G_3 and from $SO(3)$ to $E(RP^2)$ are bijective on components and induce isomorphisms on π_1, for then the corresponding maps between the classifying spaces are 2-connected. As a self homotopy equivalence of a sphere is homotopic to the identity if and only if it has degree $+1$ the inclusion of $O(3)$ into G_3 is bijective on components. The inclusion of $SO(3)$ into SG_3 gives a map betwen the evaluation fibrations of these spaces over S^2. The map of fibres from $SO(2)$ to $\Omega^2 S^2$ induces an isomorphism on π_1. On comparing the exact sequences of homotopy for these fibrations we see that the map from $SO(3)$ to SG_3 also induces an isomorphism on π_1.

The groups $SO(3)$ and $E(RP^2)$ are both connected, while $E_0(RP^2)$ has two components [Ol53]. (See also Proposition B.18 on page 145 of [Ba']). We may use obstruction theory to show that $\pi_1(E_0(RP^2))$ has order 2. Hence $\pi_1(E(RP^2))$ has order at most 4. Suppose that there were a homotopy f_t through self maps of RP^2 with $f_0 = f_1 = id_{RP^2}$ and such that the loop $f_t(*)$ is essential, where $*$ is a basepoint. Let F be the map from $RP^2 \times S^1$ to RP^2 determined by $F(p,t) = f_t(p)$, and let α and β be the generators of

$H^1(RP^2; Z/2Z)$ and $H^1(S^1; Z/2Z)$, respectively. Then $F^*\alpha = \alpha \otimes 1 + 1 \otimes \beta$ and so $(F^*\alpha)^3 = \alpha^2 \otimes \beta$ which is nonzero, contradicting $\alpha^3 = 0$. Thus there can be no such homotopy, and so the homomorphism from $\pi_1(E(RP^2))$ to $\pi_1(RP^2)$ induced by the evaluation map must be trivial. It then follows from the exact sequence of homotopy for this evaluation map that the order of $\pi_1(E(RP^2))$ is at most 2. As the composite of the maps on π_1 induced by the inclusions $SO(3) \subset E(RP^2) \subset SG_3$ is an isomorphism of groups of order 2 the first map also induces an isomorphism, and so the lemma is proven. //

Theorem 3. *Let M be a closed 4-manifold and B an aspherical closed surface. Then M is homotopy equivalent to the total space of an S^2-bundle over B if and only if $\pi_1(M) \cong \pi_1(B)$ and $\chi(M) = 2\chi(B)$.*

Proof. The conditions are clearly necessary. Suppose that they hold. Then the universal covering space $\tilde{M}$ is homotopy equivalent to S^2, since $\pi_2(M) \cong \overline{H^2(\pi; Z[\pi])} \cong Z$, by Theorem II.12, and $H_3(\tilde{M}; Z) = H_4(\tilde{M}; Z) = 0$, as $\pi_1(M)$ has one end. Let $h : M \to B$ induce an isomorphism of fundamental groups. We may assume that h is a Hurewicz fibration. As the homotopy fibre of h is $\tilde{M}$, the lemma then implies that h is fibre homotopy equivalent to the projection of an S^2-bundle over B. //

In [Me84] it is shown that the total spaces of S^2-bundles over a compact surface B are determined up to diffeomorphism by their Stiefel-Whitney classes. More precisely:

Theorem. [Me84] *Let ξ and ξ' be two S^2-bundles over a closed surface B. Then the following conditions are equivalent:*

(i) the total spaces $E(\xi)$ and $E(\xi')$ are diffeomorphic;

(ii) there is a diffeomorphism $f : B \to B$ such that $\xi = f^*\xi'$; and

(iii) $w_1(\xi) = w_1(\xi')$ if $w_1(\xi) = 0$ or $w_1(F)$, $w_1(\xi) \cup w_1(F) = w_1(\xi') \cup w_1(F)$ and $w_2(\xi) = w_2(\xi')$. //

Hence the number of diffeomorphism classes of such bundle spaces is 2 if $B = S^2$, 4 if $B = RP^2$ or if B is orientable and $\chi(B) \leq 0$, 6 if B is the Klein bottle and 8 if B is nonorientable and $\chi(B) < 0$.

Lemma 4. *Let ξ be an S^2-bundle over a closed surface B, with total space*

E and projection $p : E \to B$. Then ξ is trivial if and only if $w(E) = p^*w(B)$.

Proof. Since B is 2-dimensional, S^2-bundles over B are classified by their Stiefel-Whitney classes. On applying the Whitney product theorem and naturality to the tangent bundle of the B^3-bundle associated to ξ we find that $w(E) = p^*(w(B) \cup w(\xi))$. Thus the condition is necessary. Suppose that it holds. Then $p^*w(\xi) = 1$. As it follows from the Gysin sequence for ξ that p^* is injective, if $w(E) = p^*w(B)$ then $w(\xi) = 1$ and so ξ is trivial. //

Theorem 5. *Let M be a closed 4-manifold and B an aspherical closed surface. Then M is homotopy equivalent to the total space of an RP^2-bundle over B if and only if $\pi_1(M) \cong \pi_1(B) \times (Z/2Z)$ and $\chi(M) = \chi(B)$.*

Proof. If E is the total space of an RP^2-bundle over B, with projection p, then $\chi(E) = \chi(B)$ and the long exact sequence of homotopy gives a short exact sequence $1 \to Z/2Z \to \pi_1(E) \to \pi_1(B) \to 1$. Since the fibre has a product neighbourhood, $j^*w_1(E) = w_1(RP^2)$, where $j : RP^2 \to E$ is the inclusion of the fibre over the basepoint of B, and so $w_1(E)$ considered as a homomorphism from $\pi_1(E)$ to $Z/2Z$ splits the injection j_*. Therefore $\pi_1(E) \cong \pi_1(B) \times (Z/2Z)$ and so the conditions are necessary, as they are clearly invariant under homotopy.

Suppose that they hold, and let $w : \pi_1(M) \to Z/2Z$ be the projection onto the $Z/2Z$ factor. Then the covering space associated with the kernel of w satisfies the hypotheses of Theorem 3 and so $\tilde{M} \simeq S^2$. Therefore the homotopy fibre of the map h from M to B inducing the projection of $\pi_1(M)$ onto $\pi_1(B)$ is homotopy equivalent to RP^2. As in Theorem 3 it follows from Lemma 2 that h is fibre homotopy equivalent to the projection of an RP^2-bundle over B. //

Theorem 6. *Let M be a closed 4-manifold and $p : \hat{M} \to M$ a regular covering map, with covering group $G = Aut(p)$. If the covering space $\hat{M}$ is finitely dominated, $\pi_1(\hat{M})$ is FP_3, G has a torsion free subgroup of finite index with infinite abelianization and $H^2(G; Z[G]) \cong Z$ then M has a finite covering space which is homotopy equivalent to a closed 4-manifold which fibres over an aspherical closed surface.*

Proof. The covering space $\hat{M}$ is homotopy equivalent to a closed surface, by

Corollary A of Theorem II.9. If H is a torsion free subgroup of finite index in G which has infinite abelianization then H is also finitely presentable, and $H^2(H; Z[H]) \cong Z$. Therefore H has finite cohomological dimension, by Theorem 2.3 of [Fa74], and so is a PD_2-group, by [Fa75]. The result then follows as in Theorems 1, 3 and 5. //

Does $H^2(G; Z[G]) \cong Z$ always imply that G is virtually a surface group? Is M in fact finitely covered by such a bundle space?

Note that by Theorem II.12 and the remarks in the paragraph preceding it the total spaces of such bundles with base an aspherical surface have minimal Euler characteristic for their fundamental groups (i.e. $\chi(M) = q(\pi)$).

2. Bundles over S^2

Since S^2 is the union of two discs along a circle, an F-bundle over S^2 is determined by the homotopy class of the clutching function, which is an element of $\pi_1(Diff(F))$.

Theorem 7. *Let M be a closed 4-manifold and F a closed surface. Then M is homotopy equivalent to the total space of an F-bundle over S^2 if and only if $\chi(M) = 2\chi(F)$ and*

(i) (when $\chi(F) < 0$) there is a map $p : M \to F$ inducing an isomorphism of $\pi_1(M)$ with $\pi_1(F)$ and $p^* w(F) = w(M)$; or

(ii) (when $F = T$) $\pi_1(M) \cong Z^2$ and $w(M) = 1$, or $\pi_1(M) \cong Z \oplus (Z/nZ)$ for some $n > 0$ and, if $n = 1$ or 2, $w_1(M) = 0$; or

(iii) (when $F = K$) $\pi_1(M) \cong Z \times_{-1} Z$, $w_1(M)$ integral and nonzero, and $w_2(M) = 0$, or $\pi_1(M)$ has a presentation $< x, y | yxy^{-1} = x^{-1}, y^{2n} = 1 >$ for some $n > 0$, where $w_1(M)(x) = 0$ and $w_1(M)(y) = 1$, and there is a map $p : M \to S^2$ which induces an epimorphism on π_3; or

(iv) (when $F = S^2$) $\pi_1(M) = 1$ and the index $\sigma(M) = 0$; or

(v) (when $F = RP^2$) $\pi_1(M) = Z/2Z$, $w_1(M) \neq 0$, $w_2(M) \neq 0$ and there is a class u in $H^2(M; Z)$ which generates an infinite cyclic direct summand and has square 0.

Proof. Let $p_E : E \to S^2$ be such a bundle. Then $\chi(E) = 2\chi(F)$ and $\pi_1(E)$ is the quotient of $\pi_1(F)$ by the image of $\pi_2(S^2)$ under the connecting homomorphism ∂ of the long exact sequence of homotopy, which lies in the centre of $\pi_1(F)$. (See page 51 of [Go68]). The characteristic classes of E restrict to the characteristic classes of the fibre, as it has a product neighbourhood. As the base is 1-connected E is orientable if and only if the fibre is orientable. Thus the conditions on χ, π_1 and w_1 are all necessary. We shall treat the other assertions case by case.

(*i*) If $\chi(F) < 0$ then the component of the identity in $Diff(F)$ is contractible (see page 21 of [EE69]) and so any F-bundle over S^2 is trivial. Thus the conditions are necessary. Conversely, if they hold then p is fibre homotopy equivalent to the projection of an S^2-bundle ξ with base F, by Theorem 3. The conditions on the Stiefel-Whitney classes then imply that $w(\xi) = 1$ and hence that the bundle is trivial, by Lemma 4. Therefore M is homotopy equivalent to $S^2 \times F$.

(*ii*) If $\partial = 0$ there is a map $q : E \to T$ which induces an isomorphism of fundamental groups, and the map $(p_E, q) : E \to S^2 \times T$ is clearly a homotopy equivalence, so $w(E) = 1$. Conversely, if $\chi(M) = 0$, $\pi_1(M) \cong Z^2$ and $w(M) = 1$ then M is homotopy equivalent to $S^2 \times T$, by Theorem 3 and Lemma 4.

If $\chi(M) = 0$ and $\pi_1(M) \cong Z \oplus (Z/nZ)$ for some $n > 0$ then $\tilde{M} \simeq S^3$, and so the covering space $M_{Z/nZ}$ corresponding to the torsion subgroup Z/nZ is homotopy equivalent to a lens space L. As observed in Chapter III the manifold M is homotopy equivalent to the mapping torus of a generator of the group of covering transformations $Aut(M_{Z/nZ}/M) \cong Z$. Since the generator induces the identity on $\pi_1(L) \cong Z/nZ$ it is homotopic to id_L, if $n > 2$. This is also true if $n = 1$ or 2 and M is orientable. (See Section 29 of [Co]). Therefore M is homotopy equivalent to $L \times S^1$, which fibres over S^2 via the composition of the projection to L with the Hopf fibration of L over S^2. (Hence $w(M) = 1$ in these cases also).

(*iii*) As in part (*ii*), if $\pi_1(E) \cong Z \times_{-1} Z = \pi_1(K)$ then E is homotopy equivalent to $S^2 \times K$ and so $w_1(E) \neq 0$ while $w_2(E) = 0$. Conversely, if $\chi(M) = 0$, $\pi_1(M) \cong \pi_1(K)$, M is nonorientable but $w_1(M)$ is the reduction of an integral class and $w_2(M) = 0$ then M is homotopy equivalent to $S^2 \times K$.

Suppose now that $\partial \neq 0$. The homomorphism $\pi_3(p_E)$ induced by the bundle projection is an epimorphism. Conversely, if M satisfies these conditions and $q : M^+ \to M$ is the orientation double cover then M^+ satisfies the hypotheses of part (*ii*), and so $\tilde{M} \simeq S^3$. Therefore as $\pi_3(p)$ is onto the composition of the projection of $\tilde{M}$ onto M with p is essentially the Hopf map, and so induces isomorphisms on all higher homotopy groups. Hence the homotopy fibre of p is aspherical. As $\pi_2(M) = 0$ the fundamental group of the homotopy fibre of p is a torsion free, infinite cyclic extension of $\pi_1(M)$, and so the homotopy fibre must be K. As in Theorem 1 above the map p is fibre homotopy equivalent to a bundle projection.

(*iv*) There are just two S^2-bundles over S^2, with total spaces $S^2 \times S^2$ and $S^2 \tilde{\times} S^2 = CP^2 \sharp - CP^2$, respectively. Thus the conditions are necessary. If M satisfies these conditions then $H^2(M; Z) \cong Z^2$ and there is an element u in $H^2(M; Z)$ which generates an infinite cyclic direct summand and has square $u \cup u = 0$. Thus $u = f^* i_2$ for some map $f : M \to S^2$, where i_2 generates $H^2(S^2; Z)$, by Theorem 8.4.11 of [S]. Since u generates a direct summand there is a homology class z in $H_2(M; Z)$ such that $< u, z >= 1$, and therefore (by the Hurewicz theorem) there is a map $z : S^2 \to M$ such that fz is homotopic to id_{S^2}. The homotopy fibre of f is 1-connected and has $\pi_2 \cong Z$, by the long exact sequence of homotopy. It then follows easily from the Serre spectral sequence for f that the homotopy fibre has the homology of S^2. Therefore f is fibre homotopy equivalent to the projection of an S^2-bundle over S^2.

(*v*) Since $\pi_1(Diff(RP^2)) = Z/2Z$ (see page 21 of [EE69]) there are two RP^2-bundles over S^2. Again the conditions are clearly necessary. If they hold then $u = g^* i_2$ for some map $g : M \to S^2$. Let $q : M^+ \to M$ be the orientation double cover and $g^+ = gq$. Since $H_2(Z/2Z; Z) =$ the second homology of M is spherical. Therefore as u generates an infinite cyclic direct summand of $H^2(M; Z)$ there is a map $z = qz^+ : S^2 \to M$ such that $gz = g^+ z^+$ is homotopic to id_{S^2}. Hence the homotopy fibre of g^+ is S^2, by case (*iv*). Since the homotopy fibre of g has fundamental group $Z/2Z$ and is double covered by the homotopy fibre of g^+ it is homotopy equivalent to RP^2. It follows as in Theorem 5 that g is fibre homotopy equivalent to the projection of an RP^2-bundle over S^2. //

Theorems 1, 3 and 5 may each be rephrased as giving criteria for maps from M to B to be fibre homotopy equivalent to fibre bundle projections. With the hypotheses of Theorem 7 (and assuming also that $\partial = 0$ if $\chi(M) = 0$) we may conclude that a map $f : M \to S^2$ is fibre homotopy equivalent to a fibre bundle projection if and only if $f^* i_2$ generates an infinite cyclic direct summand of $H^2(M; Z)$.

Is there a criterion for the second case of part (*iii*) which does not refer to π_3?

We may use the classification of total spaces of S^2-bundles over surfaces of [Me84] to give examples to show that the conditions on the Stiefel-Whitney classes are independent of the other conditions when $\pi_1(M) \cong \pi_1(F)$. Note also that the nonorientable S^3- and RP^3-bundles over S^1 are not T-bundles over S^2, while $\sigma(CP^2 \sharp CP^2) \neq 0$. See Chapter IX for further information on parts (*iv*) and (*v*).

3. Bundles over RP^2.

Let $p : E \to RP^2$ be a fibre bundle with fibre an aspherical closed surface F. Since $RP^2 = Mb \cup D$ is the union of a Möbius band Mb and a disc D, the bundle p is determined by a bundle over Mb which restricts to a trivial bundle over ∂Mb, i.e. by a conjugacy class of elements of order dividing 2 in $\pi_0(Homeo(F))$, together with the class of a gluing map over $\partial Mb = \partial D$ modulo those which extend across D or Mb, i.e. an element of a quotient of $\pi_1(Homeo(F))$. Since F is aspherical $\pi_0(Homeo(F)) \cong Out(\pi_1(F))$, while $\pi_1(Homeo(F)) \cong \zeta\pi_1(F)$ [Go65]. In particular, if $\chi(F) < 0$ an F-bundle over RP^2 is determined by a conjugacy class of elements of order dividing 2 in $Out(\pi_1(F))$. (We shall consider the bundles with fibre S^2 or RP^2 in Chapter IX).

We may summarize the key properties of the algebraic invariants of such bundles in the following lemma. Let $\tilde{Z}$ be the nontrivial infinite cyclic $Z/2Z$-module and let x be the generator of $H^1(Z/2Z; \tilde{Z}) = H^1(Z/2Z; Z/2Z) = Z/2Z$. (We shall also let x denote the generator of $H^1(RP^2; \tilde{Z})$).

Lemma 8. *Let $p : E \to RP^2$ be the projection of an F-bundle, where F is an aspherical closed surface. Then* *(i)* $\chi(E) = \chi(F)$;

(ii) there is an exact sequence

$$0 \to \pi_2(E) \to Z \to \pi_1(F) \to \pi_1(E) \to Z/2Z \to 1$$

in which the connecting homomorphism $\partial : Z \to \pi_1(F)$ *has image in the centre of* $\pi_1(F)$;

(iii) if $\partial = 0$ *then* $\pi_1(E)$ *has one end and acts nontrivially on* $\pi_2(E) \cong Z$, *and the covering space* E_F *with fundamental group* $\pi_1(F)$ *is homeomorphic to* $S^2 \times F$, *so* $w_1(E)|_{\pi_1(F)} = w_1(E_F) = w_1(F)$ *(as homomorphisms from* $\pi_1(F)$ *to* $Z/2Z$*) and* $w_2(E_F) = w_1(E_F)^2$;

(iv) if $\partial \neq 0$ *then* $\chi(F) = 0$, $\pi_1(E)$ *has two ends,* $\pi_2(E) = 0$ *and* $Z/2Z$ *acts by inversion on* $\mathrm{Im}\partial$;

(v) $p^*(x)^3 = 0 \in H^3(E; p^*\tilde{Z})$.

Proof. Condition (*i*) holds since the Euler characteristic is multiplicative in fibrations, while (*ii*) is part of the long exact sequence of homotopy for p. The image of ∂ is central by [Go68], and is therefore trivial unless $\chi(F) = 0$. Conditions (*iii*) and (*iv*) then follow as the homomorphisms in this sequence are compatible with the actions of the fundamental groups, and E_F is the total space of an F-bundle over S^2, which is a trivial bundle if $\partial = 0$, by Theorem 7. Condition (*v*) holds since $H^3(RP^2; \tilde{Z}) = 0$. //

Let π be a group which is an extension of $Z/2Z$ by a normal subgroup G, and let $t \in \pi$ be an element which maps nontrivially to $\pi/G = Z/2Z$. Then $u = t^2$ is in G and conjugation by t determines an automorphism α of G which fixes u and whose square is the inner automorphism given by conjugation by u.

Conversely, let α be an automorphism of G whose square is inner, say $\alpha^2(g) = ugu^{-1}$ for all $g \in G$. Let $v = \alpha(u)$. Then $\alpha^3(g) = \alpha^2(\alpha(g)) = u\alpha(g)c^{-1} = \alpha(\alpha^2(g)) = v\alpha(g)v^{-1}$ for all $g \in G$. Therefore vu^{-1} is central. In particular, if the centre of G is trivial α fixes u, and we may define an extension

$$\xi_\alpha : 1 \to G \to \Pi_\alpha \to Z/2Z \to 1$$

in which Π_α has the presentation $< G, t_\alpha | t_\alpha g t_\alpha^{-1} = \alpha(g), t_\alpha^2 = u >$. If β is another automorphism in the same outer automorphism class then ξ_α and ξ_β are equivalent extensions. (Note that if $\beta = \alpha.c_h$, where c_h is conjugation by h, then $\beta(\alpha(h)uh) = \alpha(h)uh$ and $\beta^2(g) = \alpha(h)uh.g.(\alpha(h)uh)^{-1}$ for all $g \in G$).

Lemma 9. *If $\chi(F) < 0$ then an F-bundle over RP^2 is determined up to isomorphism by the corresponding extension of fundamental groups.*

Proof. Such bundles and extensions are each determined by an element of order 2 in $Out(\pi_1(F))$. //

Lemma 10. *Let M be a closed 4-manifold. A map $f : M \to RP^2$ is fibre homotopy equivalent to the projection of a bundle over RP^2 with fibre an aspherical closed surface if it induces an epimorphism $\pi_1(M) \to Z/2Z$ and either*

(i) $\chi(M) \leq 0$ and $f_2 : \pi_2(M) \to \pi_2(RP^2)$ is an isomorphism; or

(ii) $\chi(M) = 0$, $\pi_1(M)$ has two ends and $f_3 : \pi_3(M) \to \pi_3(RP^2)$ is an isomorphism.

Proof. Let $\tilde{M}$ be the universal covering space of M. In case (i) π is infinite, by Lemma II.14, and so $\tilde{M} \simeq S^2$, by Theorem II.11. Moreover the homotopy fibre is aspherical, and its fundamental group is a surface group. In case (ii) $\tilde{M} \simeq S^3$, by Theorem II.10, and so the lift $\tilde{f} : \tilde{M} \to S^2$ is homotopic to the Hopf map, and hence induces isomorphisms on all higher homotopy groups. Therefore the homotopy fibre of f is aspherical. As $\pi_2(M) = 0$ the fundamental group of the homotopy fibre is a (torsion free) infinite cyclic extension of π and so must be either Z^2 or $Z \tilde{\times} Z$. Thus the homotopy fibre is homotopy equivalent to a torus or Klein bottle. In both cases the argument of Theorem 1 now shows that f is fibre homotopy equivalent to a surface bundle projection. //

4. Bundles with $\partial = 0$.

In this section we shall show that the necessary conditions of Lemma 8 characterize up to homotopy equivalence the total spaces of surface bundles over RP^2 with fibre an aspherical closed surface and such that the connecting homomorphism $\partial = 0$, provided that the manifold is orientable. In the nonorientable case there is a further obstruction, of order at most 2.

Theorem 11. *Let M be a closed orientable 4-manifold. Then M is homotopy equivalent to the total space of a fibre bundle over RP^2 with fibre an aspherical closed surface and with $\partial = 0$ if and only if $\chi(M) \leq 0$, $\pi_2(M) \cong Z$, the homomorphism $p : \pi_1(M) \to Z/2Z = Aut(\pi_2(M))$ determined by the action of π on $\pi_2(M)$ is surjective, the covering space M_p corresponding to* Ker$p$ *is* a Spin-*manifold and $p^*(x)^3 = 0$.*

Proof. The conditions are clearly necessary. Suppose that they hold. Then $\chi(M_p) \leq 0$ and so Ker$p$ has infinite abelianization, by Lemma II.14. By Theorem II.11 the kernel of p is isomorphic to $\pi_1(F)$, where F is an aspherical closed surface, and so M_p is homotopy equivalent to the total space of an S^2-bundle $q : E(q) \to F$. Now $w(E(q)) = q^*(w(q) \cup w(F))$, and $w_1(q) = 0$ as $\pi_1(F)$ acts trivially on $\pi_2(M_p) \cong Z$. Since M is an orientable *Spin*-manifold and the map q is 2-connected it follows that F is orientable and $w(q) = 1$. Hence q is a trivial bundle, by Lemma 4, and so M_p is homotopy equivalent to $S^2 \times F$.

The composition of the inclusion $S^2 \vee F \subset S^2 \times F$ with a homotopy equivalence to M_p followed by the covering projection to M extends to a map $j : S^2 \vee F \vee S^1 \to M$ which induces an epimorphism of fundamental groups and an isomorphism on π_2. Up to homotopy we may assume that j is a cellular inclusion, with image in the 2-skeleton $M^{[2]}$. There is then a map $f_2 : M^{[2]} \to RP^2$ which induces the epimorphism p on fundamental groups, and such that $f_2 j|_F$ is null homotopic while $f_2 j|_{S^2}$ generates $\pi_2(RP^2)$. This map extends to a map from M to $RP^\infty = K(Z/2Z, 1)$, which we may identify with p, via the usual isomorphism $[M, K(Z/2Z, 1)] \cong Hom(\pi, Z/2Z)$.

The obstructions to the construction of a map from M to RP^2 which extends f_2 and lifts p lie in the cohomology groups $H^3(M; p^*\pi_2(RP^2))$ and $H^4(M; p^*\pi_3(RP^2))$. Now $\pi_2(RP^2) \cong \tilde{Z}$ as a $Z/2Z$-module, and so $\pi_3(RP^2) = \Gamma(\pi_2(RP^2)) \cong Z$ has the trivial $Z/2Z$-action. The first k-invariant of RP^2 is $k_1(RP^2) = x^3 \in H^3(Z/2Z; \tilde{Z})$. Since $p^*(x^3) = 0$, by assumption, there is a map $h : M \to P_2(RP^2)$ which lifts p. (In general, the lift is not unique).

Since $P_2(RP^2) \simeq RP^2 \cup (cells\ of\ dimension \geq 4)$, we may assume that h maps the 3-skeleton $M^{[3]}$ into RP^2 and $h|_{M^{[2]}} = f_2$.

The primary obstruction to a retraction of $P_2(RP^2)$ onto RP^2 is the generator t of $H^4(P_2(RP^2); Z)$. Therefore $h^*(t) \in H^4(M; Z) = H^4(M; p^*\pi_3(RP^2))$ is the single obstruction to factoring h through RP^2. (In general there is a map $h' : M \to P_2$ which factors through RP^2 and agrees with f_2 on the 2-skeleton if and only if there is a class $g \in H^2(M; p^*\tilde{Z})$ such that $g \cup g + g \cup h^*w + h^*t = 0 \in H^4(M; Z)$, where w is a generator of $H^2(P_2(RP^2); \tilde{Z})$. The groups $H^2(P_2(RP^2); \tilde{Z})$ and $H^4(P_2(RP^2); Z)$ are each infinite cyclic. See Theorem 3.24 of [Si67]).

Let $m : M_p \to M$ be the 2-fold covering and $n : \tilde{P}_2(RP^2) \to P_2(RP^2)$ be the universal covering. Then h lifts to a map $h_F : M_p \to \tilde{P}_2(RP^2)$, so that $nh_F = hm$. Since $\tilde{P}_2(RP^2)$ is a $K(Z,2)$-space, the map h_F is determined by the cohomology class $h_F^*(z)$, where z is a generator of $H^2(\tilde{P}_2(RP^2); Z) \cong Z$. Now $h_F j_F$ is a lift of $f_2 j|_F$ and so is null homotopic, while $h_F j_S$ generates $\pi_2(P_2)$, where j_F and j_S are the inclusions of the factors. Hence $h_F^*(z)$ is Poincaré dual to $j_{F*}[F]$, which has self intersection 0 in $M_p \simeq S^2 \times F$, and so $h_F^*(z)^2 = 0$. As $n^*(t)$ is a multiple of z^2, it follows that $m^*h^*(t) = 0$. If M is orientable m induces a monomorphism from $H^4(M; Z)$ to $H^4(M_p; Z)$. Thus if M is orientable there is a map $f : M \to RP^2$ which induces the epimorphism $p : \pi \to Z/2Z$ and an isomorphism on π_2. It now follows from Lemma 10 that f is fibre homotopy equivalent to the projection of an F-bundle over RP^2. //

The condition $p^*(x)^3 = 0$ is automatic if π is torsion free, as π then has cohomological dimension 2. On the other hand, if $Z/2Z$ is a (semi)direct factor of π the cohomology of $Z/2Z$ is a direct summand of that of π and so the image of x^3 in $H^3(\pi; \tilde{Z})$ is nonzero.

If M is nonorientable $H^4(M; Z) \cong Z/2Z$ and the image of $H^2(M; p^*\tilde{Z})$ under the map sending $g \in H^2(M; p^*\tilde{Z})$ to $g \cup g + g \cup h^*w$ is a subgroup. We must assume that $w_1(M_p) = w_1(F)$ (i.e., $w_1(M)|_{\pi_1(F)} = w_1(F)$ as homomorphisms from $\pi_1(F)$ to $Z/2Z$) and $w_2(M_p) = w_1(M_p)^2$, in order that M_p be homotopy equivalent to $S^2 \times F$, by Theorem 7. We then find that the class of $h^*(t)$ in the quotient group is independent of the choice of lift h and is the sole obstruction to extending f_2 to a map $f: M \to RP^2$. We do not know whether this obstruction is always 0.

5. Bundles with $\partial \neq 0$.

In this section we shall use some results from Chapters V and VIII.

If $\partial \neq 0$ then $\zeta\pi_1(F)$ is nontrivial, so F is the torus or Klein bottle, and there is a short exact sequence $0 \to \pi_1(F)/\partial Z \to \pi_1(E) \to Z/2Z \to 0$. The maximal finite normal subgroup of $\pi_1(F)/\partial Z$ is cyclic and central, and $\pi_1(E)/(\pi_1(F)/\partial Z) = Z/2Z$ acts on it by inversion, since $\pi_1(E)$ acts nontrivially on $Z = \pi_2(RP^2)$. Moreover $\tilde{E} \cong S^3 \times R$ and $\pi_1(E)$ acts trivially on $\pi_3(E)$.

If F is a torus $\pi = \pi_1(E)$ has a presentation of the form $< t, u, v | uv = vu, u^n = 1, tut^{-1} = u^{-1}, tvt^{-1} = u^a v^\epsilon, t^2 = u^b v^c >$, where $n > 0$ and $\epsilon = \pm 1$. Let τ be the maximal finite normal subgroup of π. Either

(*i*) $\tau =< u | u^n = 1 >$ is cyclic and $\pi =< t, u | u^n = 1, tut^{-1} = u^{-1} >$, so $\pi/\tau \cong Z$; or

(*ii*) $\tau =< t, u | u^n = 1, tut^{-1} = u^{-1}, t^2 = u^d >$ where either $n = 1$ and $\tau = Z/2Z$ or $n = 2d$ (since τ cannot be dihedral - see Section VIII.4). If d is odd τ is the metacyclic group $A(d, 2)$ while if $d = 2^r k$ with $r \geq 1$ and k odd τ is a generalized quaternion group $Q(2^{r+2}k)$. On replacing v by $u^{[a/2]}v$ if necessary, we may arrange that $a = 0$ (in which case $\pi \cong \tau \times Z$) or $a = 1$ (in which case $\pi =< t, u, v | u^n = 1, tut^{-1} = u^{-1}, t^2 = u^d, vtv^{-1} = tu, uv = vu >$, so $\pi/\tau \cong Z$); or

(*iii*) τ is cyclic and π has a presentation $< s, t, u | u^n = 1, sus^{-1} = tut^{-1} = u^{-1}, s^2 = u^a, t^2 = u^b >$, where $2a \equiv 2b \equiv 0 \ mod \ (n)$, so $\pi/\tau \cong D = (Z/2Z) * (Z/2Z)$.

If F is a Klein bottle π has a presentation of the form $< t, u, w | uwu^{-1} = w^{-1}, u^n = 1, tut^{-1} = u^{-1}, twt^{-1} = u^a w^\epsilon, t^2 = u^b w^c >$, where $n > 0$ is even (since $\zeta\pi_1(F)$ is generated by the square of an orientation reversing element of $\pi_1(F)$) and $\epsilon = \pm 1$. Moreover, $tw^2t^{-1} = w^{\pm 2}$ since w^2 generates the commutator subgroup of $\pi_1(F)/\partial Z$, so a is even and $2a \equiv 0 \ mod \ (n)$, $t^2 u = ut^2$ implies that $c = 0$, and $t.t^2.t^{-1} = t^2$ implies that $2b \equiv 0 \ mod \ (n)$. Therefore τ is cyclic, $\pi/\tau \cong D$ and π has a presentation of the form

(iv) $< t, u, w | uwu^{-1} = w^{-1}, u^n = 1, tut^{-1} = u^{-1}, twt^{-1} = u^a w, t^2 = u^b >$; or
(v) $< t, u, w | uwu^{-1} = w^{-1}, u^n = 1, tut^{-1} = u^{-1}, twt^{-1} = u^a w^{-1}, t^2 = u^b >$.
In all cases π has a subgroup of index at most 2 which is isomorphic to $\tau \times Z$.

Each of these groups may be realised as the fundamental group of such a bundle space. (This is most easily seen using the decomposition of the total space into a bundle over the Möbius band and a product bundle over the disc, as described in Section 3 above). In the orientable case, Ue has shown in the somewhat more general context of Seifert 4-manifolds that such bundle spaces are the total spaces of mapping tori of involutions of spherical space forms, and that if τ is not cyclic two such bundle spaces with the same group are diffeomorphic [Ue91].

Theorem 12. *Let M be a closed orientable 4-manifold such that $\chi(M) = 0$ and $\pi = \pi_1(M)$ has two ends. Then M is homotopy equivalent to the total space of a torus bundle over RP^2 if and only if π is of type (i) or (ii) above.*

Proof. Since $\chi(M) = 0$ and M is orientable $H^1(M; Z) \neq 0$; since π has two ends it follows that $\pi/\tau \cong Z$. Thus the conditions are necessary. Suppose that they hold. Then the universal covering space $\tilde{M}$ is homotopy equivalent to S^3, by Theorem II.10. In Chapter V we shall see that it is in fact homeomorphic to $R^4\backslash\{0\}$, and we shall assume this now. The infinite cyclic covering space $M^\tau = \tilde{M}/\tau$ is a PD_3-complex. Moreover the homotopy type of M is determined by π and the class of the first nonzero k-invariant, which is a generator of $H^4(\pi; Z) \cong Z/|\tau|$, modulo the action of $Aut(\pi)$.

If $\tau \cong Z/nZ$ then M^τ is homotopy equivalent to a lens space $L(n, s)$, for some s relatively prime to n. As the involution of Z/nZ which inverts a generator can be realized by an isometry of $L(n, s)$, M is homotopy equivalent to an $S^3 \times E^1$-manifold which fibres over S^1.

If $\tau \cong Q(2^{r+2}k)$ or $A(d, 2)$ then $\tau \times Z$ can only act freely and properly on $R^4\backslash\{0\}$ with the "linear" k-invariant. (For the group $A(d, 2)$, this follows from Corollary C of [HM86'], which also implies that the restriction of the k-invariant to $A(k, r + 1)$ and hence to the odd-Sylow subgroup of $Q(2^n k)$ is linear. The nonlinear k-invariants for $Q(2^n)$ have nonzero finiteness obstruction. As the k-invariants of free linear representations of $Q(2^n k)$ are given by elements in $H^4(G;Z)$ whose restrictions to Z/kZ are squares and whose restrictions to $Q(2^n)$ are squares times the basic generator (see page 120 of [Wl78]), it follows that only the linear k-invariant is realizable in this case also. Therefore M^τ is homotopy equivalent to a spherical space form S^3/τ. The class in $Out(Q(2^{r+2}k))$ represented by the automorphism which sends the generator t to tu and fixes u is induced by conjugation in $Q(2^{r+3}k)$ and so can be realized by a (fixed point free) isometry θ of $S^3/Q(2^{r+2}k)$. Hence M is homotopy equivalent to a bundle space $(S^3/Q(2^{r+2}k)) \times S^1$ or $(S^3/Q(2^{r+2}k)) \times_\theta S^1$ if $\tau \cong Q(2^{r+2}k)$. A similar conclusion holds when $\tau \cong A(d,2)$ as the corresponding automorphism is induced by conjugation in $Q(2^3d)$.

With the results of [Ue91] it follows in all cases that M is homotopy equivalent to the total space of a torus bundle over RP^2. //

The only nonorientable closed 4-manifolds with Euler characteristic 0 and fundamental group of type (*i*) or (*ii*) are homotopy equivalent to the mapping tori of the orientation reversing self homeomorphisms of S^3 and RP^3, and do not fibre over RP^2.

Theorem 12 makes no assumption that there be a homomorphism p such that $p^*(x)^3 = 0$. If τ is cyclic or $A(m,2)$ this condition is a purely algebraic consequence of the other hypotheses. For let U be a cyclic normal subgroup of maximal order in τ. (There is an unique such subgroup, except when $\tau = Q(8)$). The centralizer $C_\pi(U)$ has index 2 in π and so there is a homomorphism $p : \pi \to Z/2Z$ with kernel $C_\pi(U)$.

When τ is cyclic p factors through Z and so the induced map on cohomology factors through $H^3(Z;\tilde{Z}) = 0$.

When $\tau \cong A(m,2)$ the 2-Sylow subgroup is cyclic of order 4, and the inclusion of $Z/4Z$ into τ induces isomorphisms on cohomology with 2-local coefficients. In particular, $H^q(\tau;\tilde{Z}_{(2)}) = 0$ or $Z/2Z$ according as q is even or odd. It follows easily that the restriction from $H^3(\pi;\tilde{Z}_{(2)})$ to $H^3(Z/4Z;\tilde{Z}_{(2)})$ is an isomorphism. Let y be the image of $p^*(x)$ in $H^1(Z/4Z;\tilde{Z}_{(2)}) = Z/2Z$. Then y^2 is an element of order 2 in $H^2(Z/4Z;\tilde{Z}_{(2)}\otimes\tilde{Z}_{(2)}) = H^2(Z/4Z;Z_{(2)}) \cong Z/4Z$, and so $y^2 = 2z$ for some $z \in H^2(Z/4Z;Z_{(2)})$. But then $y^3 = 2yz = 0$ in

$H^3(Z/4Z;\tilde{Z}_{(2)}) = Z/2Z$, and so $p^*(x)^3$ has image 0 in $H^3(\pi;\tilde{Z}_{(2)}) = Z/2Z$. Since x is a 2-torsion class this implies that $p^*(x)^3 = 0$.

Is there a similar argument when τ is a generalized quaternionic group?

If M is nonorientable, $\chi(M) = 0$ and has fundamental group π of type (*iii*), (*iv*) or (*v*) above then the homotopy type of M is determined by π and the class of the k-invariant in $H^4(\pi;Z) \cong (Z/2|\tau|Z) \oplus (Z/2Z)$, modulo the action of $Aut(\pi)$, and the 2-fold covering space with fundamental group $\tau \times Z$ is homotopy equivalent to a product $L(n,s) \times S^1$. However we do not know which k-invariants correspond to the total spaces of bundles over RP^2.

CHAPTER V

SIMPLE HOMOTOPY TYPE, s-COBORDISM AND HOMEOMORPHISM

In high dimensions the problem of determining the manifolds within a given homotopy type has been successfully reduced to the determination of normal invariants and surgery obstructions. This reduction applies also in dimension 4, provided that the fundamental group is elementary amenable. Without this hypothesis we may still obtain results up to s-cobordism. Indeed, although many of the papers on surgery referred to in this chapter do not explicitly consider the 4-dimensional cases, their results may often be adapted to these cases also.

We begin by reviewing the work of Waldhausen on Mayer-Vietoris sequences for Whitehead groups. His results together with work of Farrell and Jones and Plotnick imply that for most of the surface bundles over surfaces and circle bundles over geometric 3-manifolds and for many mapping tori the Whitehead group of the fundamental group is trivial. Surgery arguments then show that 4-manifolds with torsion free, virtually poly-Z fundamental group π and Euler characteristic 0 are determined up to homeomorphism by their fundamental group. As a corollary we obtain a fibration theorem for closed 4-manifolds with torsion free, elementary amenable fundamental group. The structure sets for projective plane bundles over the torus or Klein bottle are finite.

In the next section we consider possible extensions to situations in which the fundamental group is not elementary amenable. Orientable 4-manifolds which are the total spaces of S^2-bundles over aspherical closed orientable surfaces are determined up to s-cobordism by their homotopy type. The cases considered in the final section all involve aspherical closed orientable 4-manifolds which are bundle spaces with "geometric" base and fibre. Given

Typeset by $\mathcal{A}\mathcal{M}\mathcal{S}$-TeX

such a manifold, there are only finitely many s-cobordism classes of manifolds with the same fundamental group.

1. The Whitehead group

In this section we shall rely heavily upon the work of Waldhausen in [Wd78]. A ring R is (left) *regular coherent* if every finitely presentable (left) R-module has a finite resolution by finitely generated projective R-modules, and is *regular noetherian* if moreover every finitely generated R-module is finitely presentable. A group G is regular coherent or regular noetherian if the group ring $R[G]$ is regular coherent or regular noetherian (respectively) for any regular noetherian ring R. The trivial group is regular coherent, and the class of regular coherent groups is closed under generalised free products and HNN extensions with amalgamation over regular noetherian subgroups, by Theorem 19.1 of [Wd78]. If $[G : H]$ is finite and G is torsion free then G is regular coherent if and only if H is. In particular, free groups and torsion free virtually poly-Z groups are regular coherent. Since every PD_2-group is either poly-Z or is the generalised free product of two free groups with amalgamation over infinite cyclic subgroups they are also regular coherent.

The class of groups Cl is the smallest class of groups containing the trivial group and which is closed under generalised free products and HNN extensions with amalgamation over regular coherent subgroups and under filtering direct limit. This class is also closed under taking subgroups, by Proposition 19.3 of [Wd78]. If G is in Cl then $Wh(G) = 0$, by Theorem 19.4 of [Wd78]. The argument for this theorem actually shows that if $G \cong A *_C B$ and C is regular coherent then there are "Mayer-Vietoris" sequences: $Wh(A) \oplus Wh(B) \rightarrow Wh(G) \rightarrow \tilde{K}(Z[C]) \rightarrow \tilde{K}(Z[A]) \oplus \tilde{K}(Z[B]) \rightarrow \tilde{K}(Z[G]) \rightarrow 0$, and similarly if $G \cong A*_C$. (See Sections 17.1.3 and 17.2.3 of [Wd78]).

The class Cl clearly contains all free groups, poly-Z groups and PD_2-groups. Hence homotopy equivalences between S^2-bundles over aspherical surfaces are simple. The following extension implies the corresponding result for RP^2-bundles.

Lemma 1. *Let π be a surface group. Then $Wh(\pi \times (Z/2Z)) = 0$.*

Proof. Let $\Gamma = Z[\pi]$. There is a cartesian square expressing $\Gamma[Z/2Z] = Z[\pi \times (Z/2Z)]$ as the pullback of the reduction of coefficients map from Γ to $\Gamma_2 = \Gamma/2\Gamma = Z/2Z[\pi]$ over itself. (The two maps from $\Gamma[Z/2Z]$ to Γ send the generator of $Z/2Z$ to $+1$ and -1, respectively). The Mayer-Vietoris sequence for algebraic K-theory traps $K_1(\Gamma[Z/2Z])$ between $K_2(\Gamma_2)$ and $K_1(\Gamma)^2$ (see Theorem 6.4 of [Mi71]). Now since $c.d.\pi = 2$ the higher K-theory of $R[\pi]$ can be computed in terms of the homology of π with coefficients in the K-theory of R (cf. the Corollary to Theorem 5 of the introduction of [Wd78]). In particular, the map from $K_2(\Gamma)$ to $K_2(\Gamma_2)$ is onto, while $K_1(\Gamma) = K_1(Z) \oplus (\pi/\pi')$ and $K_1(\Gamma_2) = \pi/\pi'$. It now follows easily that $K_1(\Gamma[Z/2Z])$ is generated by the images of $K_1(Z) = \{\pm 1\}$ and $\pi \times (Z/2Z)$, and so $Wh(\pi \times (Z/2Z)) = 0$. //

Lemma 1 is based on the argument for showing that $Wh(Z \oplus (Z/2Z)) = 0$ from [Kw86].

Lemma 2. *If π is an extension of $\pi_1(B)$ by $\pi_1(F)$ where B and F are aspherical closed surfaces then $Wh(\pi) = 0$.*

Proof. If $\chi(B) < 0$ then B admits a complete riemannian metric of constant negative curvature -1. Moreover the only virtually poly-Z subgroups of $\pi_1(B)$ are 1 and Z. If G is the preimage in π of such a subgroup then G is either $\pi_1(F)$ or is the group of a Haken 3-manifold. It follows easily that for any $n \geq 0$ the group $G \times Z^n$ is in Cl and so $Wh(G \times Z^n) = 0$. Therefore any such G is K-flat and so the bundle is admissible, in the terminology of [FJ86]. Hence $Wh(\pi) = 0$ by the main result of that paper.

If $\chi(B) = 0$ then this argument does not work, although if moreover $\chi(F) = 0$ then π is poly-Z so $Wh(\pi) = 0$ by [FH81]. We shall sketch an argument of Farrell for the general case. Lemma 1.4.2 and Theorem 2.1 of [FJ93] together yield a spectral sequence (with coefficients in a simplicial cosheaf) whose E^2 term is $H_i(X/\pi_1(B); Wh'_j(p^{-1}(\pi_1(B)^x)))$ and which converges to $Wh'_{i+j}(\pi)$. Here $p : \pi \to \pi_1(B)$ is the epimorphism of the extension and X is a certain universal $\pi_1(B)$-complex which is contractible and such that all the nontrivial isotropy subgroups $\pi_1(B)^x$ are infinite cyclic and the fixed point set of each infinite cyclic subgroup is a contractible (nonempty) subcomplex. The Whitehead groups with negative indices are the lower K-theory of $Z[G]$ (i.e., $Wh'_n(G) = K_n(Z[G])$ for all $n \leq -1$), while $Wh'_0(G) = \tilde{K}_0(Z[G])$ and $Wh'_1(G) = Wh(G)$. Note that $Wh'_{-n}(G)$ is a direct summand of $Wh(G \times Z^{n+1})$. If $i + j > 1$ then $Wh'_{i+j}(\pi)$ agrees rationally with the higher Whitehead group $Wh_{i+j}(\pi)$. Since the isotropy subgroups $\pi_1(B)^x$ are infinite cyclic or trivial $Wh(p^{-1}(\pi_1(B)^x) \times Z^n) = 0$ for all $n \geq 0$, by the argument of the above paragraph, and so $Wh'_j(p^{-1}(\pi_1(B)^x)) = 0$ if $j \leq 1$. Hence the spectral sequence gives $Wh(\pi) = 0$. //

A closed 3-manifold is a *Haken manifold* if it is irreducible and contains an incompressible 2-sided surface. Every Haken 3-manifold either has solvable fundamental group or may be decomposed along a finite family of disjoint incompressible tori and Klein bottles so that the complementary components are Seifert fibred or hyperbolic. It is an open question whether every closed irreducible 3-manifold with infinite fundamental group is virtually Haken (i.e., finitely covered by a Haken manifold). Every virtually Haken 3-manifold is either Haken, homotopy equivalent to a closed hyperbolic 3-manifold or Seifert-fibred [CS83]. To state the next lemma most concisely we shall say that a closed 3-manifold is *anhyperbolic* if either it has solvable fundamental group or it may be decomposed along a finite family of disjoint incompressible tori and Klein bottles so that the complementary components are Seifert fibred.

Lemma 3. *Let $\pi = \nu \times_\theta Z$ where ν is torsion free and is the fundamental group of a closed 3-manifold N which is a connected sum of anhyperbolic manifolds. Then ν is regular coherent and $Wh(\pi) = 0$.*

Proof. The group ν is a generalized free product with amalgamation along poly-Z subgroups (1, Z^2 or $Z \times_{-1} Z$) of polycyclic groups and fundamental groups of Seifert fibred 3-manifolds (possibly with boundary). The group rings of torsion free polycyclic groups are regular noetherian, and hence regular coherent. If G is the fundamental group of a Seifert fibred 3-manifold then it has a subgroup G_o of finite index which is a central extension of the fundamental group of a surface B (possibly with boundary) by Z. We may assume that G is not solvable and hence that $\chi(B) < 0$. If ∂B is nonempty then $G_o \cong Z \times F$ and so is an iterated generalized free product of copies of Z^2, with amalgamation along infinite cyclic subgroups. Otherwise we may split B along an essential curve and represent G_o as the generalised free product of two such groups, with amalgamation along a copy of Z^2. In both cases G_o is regular coherent, and therefore so is G, since $[G : G_o] < \infty$ and $c.d.G < \infty$.

Since ν is the generalised free product with amalgamation of regular coherent groups, with amalgamation along poly-Z subgroups, it is also regular coherent. Let N_i be an irreducible summand of N and let $\nu_i = \pi_1(N_i)$. If N_i is Haken then ν_i is in Cl. Otherwise N_i is a Seifert fibred 3-manifold which is not sufficiently large, and the argument of [Pl80] extends easily to show that $Wh(\nu_i \times Z^s) = 0$, for any $s \geq 0$. Since $\tilde{K}(Z[\nu_i])$ is a direct summand of $Wh(\nu_i \times Z)$, it follows that in all cases $\tilde{K}(Z[\nu_i]) = Wh(\nu_i) = 0$. The Mayer-Vietoris sequences of [Wd78] now give firstly that $Wh(\nu) = \tilde{K}(Z[\nu]) = 0$ and then that $Wh(\pi) = 0$ also. //

All 3-manifold groups are coherent as *groups*, i.e., finitely generated subgroups are finitely presentable [Sc73]. If we knew that their group *rings* were coherent then we could use [Wd78] instead of [FJ86] to give a purely algebraic proof of Lemma 2, for as surface groups are free products of free groups with amalgamation over an infinite cyclic subgroup, an extension of one surface group by another is a free product of groups with $Wh = 0$, amalgamated over the group of a surface bundle over S^1. Similarly, we could deduce from [Wd78] that $Wh(\pi_1(N) \times_\theta Z) = 0$ for N any closed 3-manifold whose irreducible factors are Haken, hyperbolic or Seifert fibred.

Lemma 4. *Let μ be a group with an infinite cyclic normal subgroup A such that $\nu = \mu/A$ is torsion free and is a free product $\nu = *_{1\leq i\leq n}\nu_i$ where each factor is the fundamental group of an irreducible 3-manifold which is Haken, hyperbolic or Seifert fibred. Then $Wh(\mu) = Wh(\nu) = 0$.*

Proof. (Note that our hypotheses allow the possibility that some of the factors ν_i are infinite cyclic). Let μ_i be the preimage of ν_i in μ, for $1 \leq i \leq n$. Then μ is the generalized free product of the μ_i's, amalgamated over infinite cyclic subgroups. For all $1 \leq i \leq n$ we have $Wh(\mu_i) = 0$, by Lemma 1.1 of [St84] if $K(\nu_i, 1)$ is Haken, by the main result of [FJ86] if it is hyperbolic, by an easy extension of the argument of [Pl80] if it is Seifert fibred but not Haken and by Theorem 19.5 of [Wd78] if ν_i is infinite cyclic. The Mayer-Vietoris sequences for the K-theory of group rings now give $Wh(\mu) = Wh(\nu) = 0$ also. //

Lemma 4 may be used to strengthen Theorem III.8 to give criteria for a closed 4-manifold M to be *simple* homotopy equivalent to the total space of an S^1-bundle, if the irreducible summands of the base N are all virtually Haken and $\pi_1(M)$ is torsion free.

2. Surgery - the elementary amenable cases

Let M be a closed 4-manifold with fundamental group π and orientation character w. The surgery obstruction maps $\sigma_{4+i}(M) : [S^iM; G/TOP] \to L^s_{4+i}(\pi, w)$ may be identified with a natural transformation from L_0-homology to L-theory. (In the nonorientable case we must use w-twisted L_0-homology). If $\pi_1(M)$ is elementary amenable and the structure set $S_{TOP}(M)$ is given the addition defined on pages 71-72 of [N] then the surgery sequence

$$[SM; G/TOP] \xrightarrow{\sigma_5} L^s_5(\pi, w) \xrightarrow{\omega} S_{TOP}(M) \xrightarrow{\eta} [M; G/TOP] \xrightarrow{\sigma_4} L^s_4(\pi, w)$$

is an exact sequence of groups [FQ]. We shall write $L_n(G, w)$ for $L^s_n(G, w)$ if $Wh(G) = 0$ and $L_n(G)$ if moreover the orientation character w is trivial. When the orientation character is nontrivial and otherwise clear from the context we shall write $L_n(G, -)$.

Theorem 5. *Let M be a closed 4-manifold with $\chi(M) = 0$ and with $\pi = \pi_1(M)$ a torsion free virtually poly-Z group of Hirsch length 4. Then M is determined up to homeomorphism by π.*

Proof. Since π is torsion free and solvable it has a safe extension, by Theorem I.5, and since $h(\pi) > 2$ it has one end and $H^2(\pi; Z[\pi]) = 0$, by an LHSSS argument. Therefore M is aspherical, by the Corollary to Theorem II.7. As the surgery obstruction homomorphisms are isomorphisms, by Corollary B of [FJ88], the theorem now follows. //

We shall weaken the hypotheses on this theorem considerably in Chapter VI.

Given $\alpha : S^2 \to M$, let $\beta : S^4 \to M$ be the composition $\alpha\eta S\eta$, where η is the Hopf map, and let $s : M \to M \vee S^4$ be the pinch map obtained by shrinking the boundary of a 4-disc in M. Then the composite $f_\alpha = (id_E, \beta)s$ is a self homotopy equivalence of M.

Lemma 6. *Let M be a closed 4-manifold and let $\alpha : S^2 \to M$ be a map such that $\alpha_*[S^2] \neq 0$ in $H_2(M; Z/2Z)$ and $\alpha^* w_2(M) = 0$. Then the self homotopy equivalence f_α has nontrivial normal invariant in $[M; G/TOP]$.*

Proof. Since $\alpha_*[S^2] \neq 0$ there is a map $y : Y \to M$ with domain a closed surface which is transverse to f_α and such that $\alpha_*[S^2].y_*[Y] = 1$ modulo (2). Then the Arf invariant of the induced normal map over Y is nontrivial, and so f_α cannot be normally cobordant to a homeomorphism. (Compare the argument of Theorem 5.1 of [CH]). //

In the next lemma we do not assume that the fundamental group is elementary amenable.

Lemma 7. *Let M be a closed 4-manifold such that $\pi = \pi_1(M)$ is the group of a finite graph of groups, all of whose vertex groups are infinite cyclic. Then the surgery obstruction maps $\sigma_4(M)$ and $\sigma_5(M)$ are epimorphisms.*

Proof. Since $c.d.\pi \leq 2$ the homomorphism $c_{M*} : H_*(M; D) \to H_*(\pi; D)$ is onto for any local coefficient module D. Since π is in Cl we have $Wh(\pi) = 0$ and a comparison of Mayer-Vietoris sequences gives an isomorphism from $H_*(\pi; L_0^w)$ to $L_*(\pi, w)$ [St87]. We may now apply Formula 1.10 of [HMTW] to compute the surgery obstruction. (This formula involves localization at 2. However the low dimensional characteristic classes needed for the lemma are integral). //

The class of groups considered in this lemma includes the PD_2-groups. Note however that if π is such a group w need not be the canonical orientation character. We can only use the result on σ_5 directly if π is elementary amenable. See Theorem 12 and its corollary below for a substitute.

Theorem 8. *Let M be a closed 4-manifold with $\chi(M) = 0$ and $\pi = \pi_1(M) \cong Z^2$ or $Z \times_{-1} Z$. Then M is homeomorphic to the total space of an S^2-bundle* *over the torus or Klein bottle.*

Proof. There is a homotopy equivalence $f : M \to E$, by Theorem IV.2, which must be simple, since $Wh(\pi) = 0$. The surgery obstruction map $\sigma_4(E)$ is onto, by Lemma 7, and so there are at most two normal cobordism classes of homotopy equivalences $h : X \to E$. Let $j : S^2 \to E$ be the inclusion of a fibre. Then $j_*[S^2]$ generates the kernel of the natural homomorphism from $H_2(E; Z/2Z) \cong (Z/2Z)^2$ to $H_2(\pi; Z/2Z) \cong Z/2Z$ and so is nonzero, while $w_2(E)(j_*[S^2]) = j^* w_2(E) = 0$. Therefore f_j has nontrivial normal invariant, by Lemma 6, and so each of these two normal cobordism classes contains a self homotopy equivalence of E. Therefore there is a normal cobordism $F : V \to E \times [0, 1]$ from f to some self homotopy equivalence h of E. Since $\sigma_5(E)$ is onto it follows from the exact sequence of surgery that f is homotopic to a homeomorphism. //

We may now give an analogue of the Farrell and Stallings fibration theorems for 4-manifolds with torsion free elementary amenable fundamental group.

Theorem 9. *Let M be a closed 4-manifold such that $\pi = \pi_1(M)$ is a torsion free elementary amenable group. A map $f : M \to S^1$ is homotopic to a fibre bundle projection if and only if $\chi(M) = 0$ and f induces an epimorphism from π to Z with almost finitely presentable kernel.*

Proof. The conditions are clearly necessary. Suppose that they hold. Let $\nu = \mathrm{Ker}\pi_1(f)$, let M_ν be the infinite cyclic covering space of M with fundamental group ν and let $t : M_\nu \to M_\nu$ be a generator of the group of covering transformations. By the Corollary to Theorem III.4 either $\nu = 1$ (so $M_\nu \simeq S^3$) or $\nu \cong Z$ (so $M_\nu \simeq S^2 \times S^1$ or $S^2 \tilde{\times} S^1$) or M is aspherical. In the latter case π is a torsion free virtually poly-Z group, by Theorem I.5 and Theorem 9.23 of [Bi]. Thus in all cases there is a homotopy equivalence f_ν from M_ν to a closed 3-manifold N. Moreover the self homotopy equivalence $f_\nu t f_\nu^{-1}$ of N is homotopic to a homeomorphism, g say, and so f is fibre homotopy equivalent to the canonical projection of the mapping torus $M(g)$ onto S^1. A simple surgery argument (using Proposition 11.6A of [FQ] and the fact that $L_5(Z, -) = 0$, Theorem 8 and Theorem 5, respectively) now shows that any homotopy equivalence from M to $M(g)$ is homotopic to a homeomorphism.//

It is easy to show that the structure sets of the RP^2-bundles over the torus or Klein bottle are finite.

Theorem 10. *Let M be a closed connected 4-manifold with $\chi(M) = 0$ and with fundamental group π. If π is virtually Z^2 then $S_{TOP}(M)$ has order at most 32.*

Proof. By Theorem IV.5 we may assume that M is the total space of an RP^2-bundle $p : M \to B$ with base a torus or Klein bottle. Then $L_5(\pi, w_1) \cong L_4(Z/2Z, -)^2 \cong (Z/2Z)^2$ and $L_4(\pi, w_1) \cong L_4(Z/2Z, -) \oplus L_2(Z/2Z, -) \cong (Z/2Z)^2$, by Theorem 12.6 of [W]. As M is nonorientable $H^4(M; Z) = Z/2Z$ and as $H^1(M; Z/2Z) = (Z/2Z)^3$ and $\chi(M) = 0$ we have $H^2(M; Z/2Z) \cong (Z/2Z)^4$. As $[M; G/TOP] = H^4(M; Z) \oplus H^2(M; Z/2Z)$ this is enough to show that $S_{TOP}(M)$ has order at most 128.

The argument of the splitting theorem quoted in the above paragraph shows also that the surgery obstructions are determined by the Kervaire-Arf invariants of the surgery problems induced over submanifolds of codimension ≤ 2. The Kervaire-Arf invariant of a 4-dimensional normal map represented by $g : M \to G/TOP$ is the image of $\sigma_4(M)(g)$ under the projection onto $L_4(Z/2Z, -) \cong Z/2Z$ and is given by $c(g) = w_1(M)^2 \cup g^*\kappa_2[M]$, by Theorem 13B.5 of [W]. As every element of $H^2(M; Z/2Z)$ is equal to $g^*\kappa_2$ for some such g and as $w_1(M)^2 \neq 0$ this composite is onto. Similarly there is a normal map $f_2 : X_2 \to RP^2$, which we may assume is a homeomorphism over a disc $\Delta \subset RP^2$, with $c(f_2) \neq 0$ in $L_2(Z/2Z, -)$. If $M = RP^2 \times B$ then $f_2 \times id_B : X_2 \times B \to RP^2 \times B$ is a normal map with surgery obstruction $(0, c(f_2)) \in L_4(Z/2Z, -) \oplus L_2(Z/2Z, -)$. The two nontrivial bundles may be obtained from the product bundles by cutting M along $RP^2 \times \partial\Delta$ and regluing via the twist map of $RP^2 \times S^1$. As $f|_{f^{-1}(\Delta)}$ is a homeomorphism these normal maps may be compatibly modified. It follows that in all cases $\sigma_4(M)$ is onto and so $S_{TOP}(M)$ has order at most 32. //

In the case of RP^2-bundles over the torus we may obtain a more precise result.

Theorem 11. *Let M be a closed connected 4-manifold with $\chi(M) = 0$ and* *with fundamental group* $\pi \cong Z^2 \oplus (Z/2Z)$. *Then* $S_{TOP}(M)$ *has order 8.*

Proof. Recall that $w_1(M)$ maps the torsion subgroup of π onto $Z/2Z$, by Theorem IV.4. Let $\lambda_1, \lambda_2 : \pi \to Z$ be epimorphisms forming a basis for $Hom(\pi, Z)$ and let $t_1, t_2 \in \mathrm{Ker} w_1$ represent a dual basis for $\pi/(torsion)$ (i.e., $\lambda_i(t_j) = \delta_{ij}$). Then $p_i(g) = g - \lambda_i(g)t_i$ for all $g \in \pi$ defines a splitting for the inclusion $k_i : \mathrm{Ker}\lambda_i \to \pi$, and $p_i k_{3-i}$ factors through $Z/2Z$, for $i = 1, 2$. As the maps p_i and k_i are compatible with the orientation character and as $L_5(Z/2Z, -) = 0$ it follows that $L_5(k_1)$ and $L_5(k_2)$ are inclusions of complementary summands of $L_5(\pi, w_1) \cong (Z/2Z)^2$, split by the projections $L_5(p_1)$ and $L_5(p_2)$.

Let γ_i be a simple closed curve in B which represents $t_i \in \pi$. Then γ_i has a product neighbourhood $N_i \cong S^1 \times [-1, 1]$ and $U_i = p^{-1}(N_i)$ is homeomorphic to $RP^2 \times S^1 \times [-1, 1]$. There is a normal map $f_4 : X_4 \to RP^2 \times [-1, 1]^2$ (*rel* boundary) with $c(f_4) \neq 0$ in $L_4(Z/2Z, -)$. Let $Y_i = (M \backslash intU_i) \times [-1, 1] \cup X_4 \times S^1$, where we identify $(\partial U_i) \times [-1, 1] = RP^2 \times S^1 \times S^0 \times [-1, 1]$ with $RP^2 \times [-1, 1] \times S^0 \times S^1$ in $\partial X_4 \times S^1$. If we match together $id_{(M \backslash intU_i) \times [-1,1]}$ and $f_4 \times id_{S^1}$ we obtain a normal cobordism Q_i from id_M to itself. The image of the surgery obstruction $\sigma_5(M)(Q_i)$ in $L_4(\mathrm{Ker}\lambda_i, w_1) \cong L_4(Z/2Z, -)$ under the splitting homomorphism is $c_4(f)$. On the other hand its image in $L_4(\mathrm{Ker}\lambda_{3-i}, w_1)$ is 0, and so it generates the image of $L_5(k_{3-i})$. Thus $L_5(\pi, w_1)$ is generated by $\sigma_5(M)(Q_1)$ and $\sigma_5(M)(Q_2)$, and so acts trivially on the class of id_M. Since the surgery sequence $[SM; G/TOP] \to L_5(\pi, w_1) \to S_{TOP}(M)$ is an exact sequence of groups $S_{TOP}(M)$ has order 8. //

Does a similar idea work to show that $L_5(\pi, w_1)$ acts trivially on each class in $S_{TOP}(M)$ when M is an RP^2-bundle over the Klein bottle? If so, then $S_{TOP}(M)$ has order 8 in these cases also. Are these manifolds determined up to homeomorphism by their homotopy type?

We shall defer discussion of 4-manifolds with universal covering space homotopy equivalent to S^3 and bundles with base and fibre surfaces of positive Euler characteristic until Chapters VIII and IX, respectively.

3. S^2-bundles over other aspherical surfaces

As it is not yet known whether 5-dimensional s-cobordisms over other fundamental groups are products, we shall redefine the structure set by setting $S^s_{TOP}(M) = \{f : N \to M | N$ *a closed TOP 4-manifold, f a simple homotopy equivalence*$\}/\approx$, where $f_1 \approx f_2$ if there is a map $F : W \to M$ with domain W an s-cobordism with $\partial W = N_1 \cup N_2$ and $F|_{N_i} = f_i$ for $i = 1, 2$. If the s-cobordism theorem holds over $\pi_1(M)$ this is the usual TOP structure set for M. In general $\sigma_4(M)$ is trivial on the image of $S^s_{TOP}(M)$, but we do not know whether a 4-dimensional normal map with trivial surgery obstruction must be normally cobordant to a simple homotopy equivalence. In our applications we shall always have a simple homotopy equivalence in hand, and so if we can show that σ_4 is injective we can conclude that the homotopy equivalence is normally cobordant to the identity. A more serious problem is that it is not clear how to define the action ω. We shall be able to circumvent this problem by *ad hoc* arguments in some cases, in which the elements of $L_5(\pi)$ are detected by their images under projections of π onto Z.

Theorem 12. *Let M be a closed orientable 4-manifold with fundamental group π and let $\lambda_* : L_5(\pi) \to L_5(Z)^d = Z^d$ be the homomorphism induced by a basis $\{\lambda_1, ..., \lambda_d\}$ for $Hom(\pi, Z)$. Given any simple homotopy equivalence $f : M \to M_1$ and element $\theta \in L_5(Z)^d$ there is a normal cobordism from f to itself whose surgery obstruction in $L_5(\pi)$ has image θ under λ_*.*

Proof. Let $\{\gamma_1, ..., \gamma_d\}$ be simple closed curves in M such that $\lambda_i([\gamma_j]) = \delta_{ij}$ for $1 \leq i, j \leq d$. Since M is orientable γ_i has a product neighbourhood $U_i \cong S^1 \times D^3$. Let P be the E_8 manifold [FQ] and delete the interior of a submanifold homeomorphic to $D^3 \times [0, 1]$ to obtain P_o. There is a normal map $p : P_o \to D^3 \times [0, 1]$ (*rel* boundary). The surgery obstruction $\sigma(p)$ is a generator of $L_4(Z) \cong Z$. Let $Y_i = (M \backslash intU_i) \times [0, 1] \cup P_o \times S^1$, where we identify $(\partial U_i) \times [0, 1] = S^1 \times S^2 \times [0, 1]$ with $S^2 \times [0, 1] \times S^1$ in $\partial P_o \times S^1$. If we match together $f|_{(M \backslash intU_i)} \times id_{[0,1]}$ and $(f \times id_{[0,1]}) \circ (p \times id_{S^1})$ we obtain a normal cobordism Q_i from id_M to itself. The image of the surgery obstruction $\sigma_5(M)(Q_i)$ in $L_5(Z) \cong L_4(1) = Z$ under $L_5(\lambda_j)$ is given by a codimension-1 signature, and generates $L_5(Z)$ if $j = i$ and is 0 otherwise. The theorem now follows from the additivity of surgery obstructions. //

If π is free or is a PD_2-group of orientable type then the homomorphism λ_* is an isomorphism [Ca]. In most of the other cases of interest to us the following corollary applies.

Corollary. *If $\sigma_5(M)$ is an epimorphism then $S^s_{TOP}(M)$ is finite.*

Proof. Since the signature difference maps $[M; G/TOP] = H^4(M; Z) \oplus H^2(M; Z/2Z)$ onto $L_4(1) = Z$ there are only finitely many normal cobordism classes of simple homotopy equivalences $f : M_1 \to M$. Suppose that $F : N \to M \times I$ is a normal cobordism between two simple homotopy equivalences $F_- = F|\partial_- N$ and $F_+ = F|\partial_+ N$. By Theorem 12 there is another normal cobordism $F' : N' \to M \times I$ from F_+ to itself with $\lambda_*(\sigma_5(M)(F')) = \lambda_*(-\sigma_5(M)(F))$. The union of these two normal cobordisms along $\partial_+ N = \partial_- N'$ is a normal cobordism from F_- to F_+ with surgery obstruction in $\mathrm{Ker}\lambda_*$. Since $\sigma_5(M)$ and λ_* are epimorphisms this kernel is finite. If this obstruction is 0 we may obtain an s-cobordism W by 5-dimensional surgery (rel ∂). //

Theorem 13. *Let E be an orientable 4-manifold which is the total space of an S^2-bundle over an aspherical closed surface B. Then $S^s_{TOP}(E)$ has at most 2 elements, and has exactly one element if B is orientable.*

Proof. Let $\pi = \pi_1(E) \cong \pi_1(B)$ and let $f : M \to E$ be a homotopy equivalence. Then f is simple, since $Wh(\pi) = 0$. The surgery obstruction map $\sigma_4(E)$ is onto, by Lemma 7, and so there are at most two normal cobordism classes of homotopy equivalences $h : X \to E$. Let $j : S^2 \to E$ be the inclusion of a fibre. Then $j_*[S^2]$ generates the kernel of the natural homomorphism from $H_2(E; Z/2Z) \cong (Z/2Z)^2$ to $H_2(\pi; Z/2Z) \cong Z/2Z$ and so is nonzero, while $w_2(E)(j_*[S^2]) = j^* w_2(E) = 0$. Therefore f_j has nontrivial normal invariant, by Lemma 6, and so each of these two normal cobordism classes contains a self homotopy equivalence of E. Therefore there is a normal cobordism $F : V \to E \times [0,1]$ from f to some self homotopy equivalence h of E. As the group $L_5(\pi)$ is isomorphic to π/π' [Ca] the theorem now follows from Theorem 12 and its corollary. //

Does this theorem remain true without the hypotheses that E and B be orientable?

Corollary. *If M is homotopy equivalent to the total space of an RP^2- or S^2-bundle over a closed aspherical surface then $\tilde{M}$ is homeomorphic to $S^2 \times R^2$.*

Proof. We may assume that M is orientable and π is of orientable type. By the theorem there is an s-cobordism W from M to a bundle space E. The universal covering space $\tilde{W}$ is a proper s-cobordism from $\tilde{M}$ to $\tilde{E} \cong S^2 \times R^2$. Since the end of $\tilde{E}$ is tame and has fundamental group Z we may apply Corollary 7.3B of [FQ] to conclude that $\tilde{W}$ is homeomorphic to a product. Hence $\tilde{M}$ is homeomorphic to $S^2 \times R^2$. //

4. Aspherical bundle spaces

It remains an open question whether aspherical closed manifolds with isomorphic fundamental groups must be homeomorphic. This has been verified in higher dimensions in many cases, in particular under geometric assumptions [FJ], and under assumptions on the combinatorial structure of the group [Ca73]. In this section we shall apply such results to show that the structure sets of aspherical orientable 4-dimensional bundle spaces with base or fibre a closed surface or a closed Seifert fibred 3-manifold or a Sol^3-manifold are finite. As the universal covering space of such a bundle is homeomorphic to R^4, the fundamental group is 1-connected at ∞, and so any homotopy equivalent 4-manifold is also covered by R^4. (There are aspherical 4-manifolds whose universal covering space is not 1-connected at ∞ [Da83]).

Theorem 14. *Let E be the total space of an F-bundle over B where B and F are aspherical closed surfaces, and let $\pi = \pi_1(E)$ and $w = w_1(E)$. Then the surgery obstruction maps $\sigma_4(E)$ and $\sigma_5(E)$ are isomorphisms.*

Proof. Since $\pi_1(B)$ is either an HNN extension of Z or a generalised free product $F *_Z F'$, where F and F' are free groups, π is a square root closed generalised free product with amalgamation of groups in Cl. Since E is aspherical the lemma follows on comparison of the Mayer-Vietoris sequences for L_0-homology and L-theory, as in Proposition 2.6 of [St84]. (Note that even when $\chi(B) = 0$ the groups arising in intermediate stages of the argument all have trivial Whitehead groups). //

Corollary. *Let E be an orientable 4-manifold which is the total space of an F-bundle over B where B and F are aspherical closed surfaces. Then $S^s_{TOP}(E)$ is finite.* //

As the characterization of fundamental groups of Seifert fibred 3-manifolds given in [Hi85] and used in each of the next two theorems has a gap, we have given a correct proof as Theorem 6 of the Appendix.

Theorem 15. *A closed 4-manifold M is simple homotopy equivalent to the mapping torus of a self homeomorphism of an aspherical closed 3-manifold which is Seifert fibred or admits a Sol^3-structure if and only if $\chi(M) = 0$ and $\pi = \pi_1(M)$ is an extension of Z by an almost finitely presentable normal subgroup ν with one end and which has subgroups $A < B$ such that A is a nontrivial finitely generated free abelian group which is normal in π and B has finite index in ν and infinite abelianization. If A has rank at least 2 then M is homeomorphic to such a mapping torus. If M is orientable the structure set $S^s_{TOP}(M)$ is finite.*

Proof. The conditions are clearly necessary. If they hold then $H^s(\pi; Z[\pi]) = 0$ for $s \leq 2$ and $Z[\pi]$ has a safe extension, so M is aspherical, by the Corollary to Theorem II.7. Hence $c.d.\nu = 3$, as $c.d.\nu < c.d.\pi$, by [St77], and $c.d.\nu + c.d.Z \geq c.d.\pi = 4$, by Theorem 5.6 of [Bi]. If $A \cong Z$ then ν is a PD_3-group, by Theorem III.4, and hence is the group of a closed Seifert fibred 3-manifold of type Nil^3, $\tilde{SL}$ or $H^2 \times E^1$ [Hi85]. If $A \cong Z^2$ then ν/A is virtually free, by Theorem 8.4 of [Bi]. But then π/A has virtually finite cohomological dimension. Hence it is virtually a surface group, by Theorem 9.11 of [Bi] (with an LHSSS corner argument to identify the dualizing module). As it has a finitely generated infinite normal subgroup of infinite index it must be poly-Z and so π is virtually poly-Z. If $A \cong Z^3$ then $[\nu : A]$ is finite, by Theorem 8.2 of [Bi], and so ν is the fundamental group of a flat 3-manifold.

In all cases ν is isomorphic to the fundamental group of an aspherical closed 3-manifold N which is either Seifert fibred or a Sol^3-manifold, and the outer automorphism class $[\theta]$ determined by the extension may be realised by a self homeomorphism Θ of N. The manifold M is homotopy equivalent to the mapping torus $M(\Theta)$. Since $Wh(\pi) = 0$, by Lemma 3, any such homotopy equivalence is simple.

If ν is solvable then π is virtually poly-Z, and so M is homeomorphic to $M(\Theta)$, by Theorem 5. Otherwise N is a closed $\tilde{SL}$- or $H^2 \times E^1$-manifold. Since N has a metric of nonpositive sectional curvature the surgery obstruction homomorphisms $\sigma_i(N) : [S^{3-i}N; G/TOP] \to L_i(\nu, w_1(N))$ are all isomorphisms [FJ90]. Comparison of the Mayer-Vietoris sequences for L_0-homology and L-theory (as in Proposition 2.6 of [St84]) shows that $\sigma_4(M)$ and $\sigma_5(M)$ are also isomorphisms. Hence if M is orientable $S^s_{TOP}(M)$ is finite, by the corollary to Theorem 12. //

If every PD_3-group is a 3-manifold group and the geometrization conjecture for atoroidal 3-manifolds is true then the fundamental groups of closed hyperbolic 3-manifolds may be characterized as PD_3-groups having no noncyclic abelian subgroup. Assuming this, and assuming also that group rings of such hyperbolic groups are regular coherent, Theorem 15 could be extended to show that a closed 4-manifold M with fundamental group π is simple homotopy equivalent to the mapping torus of a self homeomorphism of a hyperbolic 3-manifold if and only if $\chi(M) = 0$ and π is an extension of Z by an almost finitely presentable normal subgroup ν with one end and which has no noncyclic abelian subgroup, and has a subgroup of finite index which is isomorphic to $Z \times \nu$.

There is a parallel characterization of total spaces of circle bundles over aspherical Seifert fibred 3-manifolds and Sol^3-manifolds.

Theorem 16. *A closed 4-manifold M is simple homotopy equivalent to the total space of an S^1-bundle over an aspherical closed 3-manifold which is Seifert fibred or admits a Sol^3-structure if and only if $\chi(M) = 0$ and $\pi = \pi_1(M)$ has normal subgroups $A < B < C$ such that A is infinite cyclic, π/A has one end and finite cohomological dimension, B/A is nontrivial abelian, C has finite index in π and C/A has infinite abelianization. If B/A has rank at least 2 then M is homeomorphic to such a bundle space. If M is orientable* *the structure set $S^s_{TOP}(M)$ is finite.*

Proof. The conditions are clearly necessary. If they hold then π/A is a PD_3-group by Theorem III.9 and has a subgroup of finite index with a nontrivial abelian normal subgroup and infinite abelianization. Therefore π/A is the group of a closed Seifert fibred 3-manifold (if there is such a subgroup B with B/A of rank 1 or 3) or of a closed 3-manifold with a Sol-structure (if the rank of B/A is 2, for every such subgroup B) [Hi85]. Hence M is homotopy equivalent to the total space E of such a bundle. Any such homotopy equivalence must be simple, by Lemma 4.

If there is such a subgroup B such that B/A has rank at least 2 (corresponding to the base having an E^3-, Nil^3- or Sol^3-structure) then π is torsion free and virtually poly-Z, so M is homeomorphic to such a bundle space by Theorem 5. If M is orientable the surgery obstruction maps $\theta_4(M)$ and $\theta_5(M)$ are isomorphisms [NS85], and so $S^s_{TOP}(M)$ is finite, by the Corollary to Theorem 12. //

If the above conjectural characterization of fundamental groups of closed hyperbolic 3-manifolds is valid then we could use the argument of Theorem 10.7 of [FJ89] instead of [NS85] to extend Theorem 16 to show that a closed 4-manifold M with fundamental group π is simple homotopy equivalent to the total space of an S^1-bundle over a hyperbolic 3-manifold if and only if $\chi(M) = 0$ and π has an infinite cyclic normal subgroup A such that π/A has one end and finite cohomological dimension and has no noncyclic abelian subgroup, and that the corresponding structure set is finite. We may similarly obtain the latter conclusion when the base is Haken with square root closed accessible fundamental group, using [Ca73] instead of [NS85]. However we do not yet have good intrinsic characterizations of the fundamental groups of such 3-manifolds comparable to the result of [Hi85].

CHAPTER VI

ASPHERICAL GEOMETRIES

An *n-dimensional geometry* in the sense of Thurston is a pair (X, G_X) where X is a complete 1-connected n-dimensional Riemannian manifold and G_X is a group of isometries which acts transitively on X and has discrete subgroups Γ such that $\Gamma\backslash X$ has finite volume. Using an equivalent formulation, we may say that a closed manifold M is an *X-manifold*, or *admits a geometry of type X*, if it is homeomorphic to a quotient $\Gamma\backslash X$ for some torsion free discrete group Γ of isometries of a 1-connected homogeneous space $X = G/K$, where G is a connected Lie group and K is a compact subgroup of G such that the intersection of the conjugates of K is trivial, and X has a G-invariant metric. If G is solvable we shall say that the geometry is of *solvable Lie type*.

Thurston showed that there are eight 3-dimensional geometries (E^3, Nil^3, Sol^3, $\tilde{SL}$, $H^2 \times E^1$, H^3, $S^2 \times E^1$ and S^3). The first five of these are realized by aspherical Seifert fibred 3-manifolds; such manifolds are determined among irreducible 3-manifolds by their fundamental groups. The possible groups are the PD_3-groups with nontrivial Hirsch-Plotkin radical and which have subgroups of finite index with infinite abelianization. There are just four $S^2 \times E^1$-manifolds. The determination of the closed 3-manifolds admitting hyperbolic or spherical geometries remains incomplete; in particular, it is not yet known whether every aspherical 3-manifold whose fundamental group contains no rank 2 abelian subgroup must be hyperbolic nor whether every 3-manifold with finite fundamental group must be spherical.

There are 19 maximal 4-dimensional geometries; one of these is in fact a countably infinite family of closely related geometries, and one is not realizable by any closed manifold [Fi]. The geometry is determined by the homotopy type, and indeed by the fundamental group [Wl86, Ko92]. In 13 cases the model X is homeomorphic to R^4, and so manifolds admitting such geometries

Typeset by $\mathcal{A}\mathcal{M}\mathcal{S}$-TEX

are aspherical. Six of these geometries (E^4, Nil^4, $Nil^3 \times E^1$, $Sol^4_{m,n}$, Sol^4_0 and Sol^4_1) are realized by infrasolvmanifolds; we shall use topological surgery to show such manifolds may be characterized up to homeomorphism by the algebraic conditions that the Euler characteristic is 0 and the fundamental group has a locally nilpotent normal subgroup of Hirsch length at least 3. The fundamental group is then a torsion free virtually poly-Z group of Hirsch length 4. Conversely, every such group is the fundamental group of such a manifold, which is determined up to homeomorphism by the group. Closed manifolds admitting one of the other geometries of aspherical type ($H^3 \times E^1$, $\tilde{SL} \times E^1$, $H^2 \times E^2$, $H^2 \times H^2$, H^4 and $H^2(C)$) may be characterised up to simple homotopy equivalence by their fundamental group and Euler characteristic. However it is unknown to what extent surgery arguments apply in these cases. Moreover we do not yet have good characterizations of the possible fundamental groups.

The six remaining geometries are all of nonaspherical type. Three of these geometries ($S^2 \times E^2$, $S^2 \times H^2$ and $S^3 \times E^1$) have models homeomorphic to $S^2 \times R^2$ or $S^3 \times R$. (Note that we shall use E^n or H^n to refer to the *geometry* and R^n to refer to the underlying *topological space*). The remaining three (S^4, CP^2 and $S^2 \times S^2$) have compact models, and there are only eleven such manifolds. We shall discuss these nonaspherical geometries in Chapters VII, VIII and IX.

1. Infrasolvmanifolds

An *infrasolvmanifold* is a quotient $\pi\backslash G/K$ where G is a Lie group whose maximal connected subgroup G_o is solvable and of finite index in G, K is a maximal compact subgroup of G which meets each component of G and π is a torsion free discrete cocompact subgroup of G. An infrasolvmanifold M is aspherical since $G/K = G_o/K \cap G_o$ is homeomorphic to R^m [Iw49] and so its fundamental group π is a PD_m group, where m is the dimension of M; since π is also virtually solvable it is thus virtually poly-Z of Hirsch length m, by Theorem 9.23 of [Bi]; moreover $\chi(M) = 0$. As K is compact we may equip $X = G/K$ with a G-invariant metric, and then π acts isometrically, so infrasolvmanifolds have geometric structures. Examination of Filipkiewicz's list shows that the geometry must be one of the six (families of) geometries "of solvable Lie type": $E^4, Nil^3 \times E^1, Nil^4, Sol^4_{m,n}, Sol^4_0$ or Sol^4_1 and that, conversely, a closed manifold with a geometry of solvable Lie type is an infrasolvmanifold.

The group π is a discrete cocompact subgroup of a *connected* solvable Lie group G if and only if $\pi/\sqrt{\pi}$ is abelian, while if G is simply connected $\pi/\sqrt{\pi}$ must also be torsion free. (See Sections 4.29-31 of [Rg]).

Lemma 1. *Let π be a torsion free virtually poly-Z group of Hirsch length 4. Then $h(\sqrt{\pi}) \geq 3$.*

Proof. Let S be a solvable normal subgroup of finite index in π. Then the lowest nontrivial term of the derived series of S is an abelian subgroup which is characteristic in S and so normal in π. Hence $\sqrt{\pi} \neq 1$. If $h(\sqrt{\pi}) \leq 2$ then $\sqrt{\pi} \cong Z$ or Z^2. Suppose π has an infinite cyclic normal subgroup A. On replacing π by a normal subgroup σ of finite index we may assume that A is central and that σ/A is poly-Z. Let B be the preimage in σ of a nontrivial abelian normal subgroup of σ/A. Then B is nilpotent (since A is central and B/A is abelian) and $h(B) > 1$ (since $B/A \neq 1$ and σ/A is torsion free). Hence $h(\sqrt{\pi}) \geq h(\sqrt{\sigma}) > 1$.

Suppose now that π has a normal subgroup N isomorphic to Z^2. Since $Aut(N) \cong GL(2,Z)$ is virtually free the kernel of the natural map from π to $Aut(N)$ is nontrivial, so $h(C_\pi(N)) \geq 3$. Since $h(\pi/N) = 2$ the quotient π/N is virtually abelian, and so $C_\pi(N)$ is virtually nilpotent.

In all cases we must have $h(\sqrt{\pi}) \geq 3$. //

We shall describe the finitely generated, torsion free nilpotent groups of Hirsch length at most 4 more explicitly in the next section.

Theorem 2. *Let M be a closed 4-manifold with fundamental group π. The following conditions are equivalent:*

(i) M is homeomorphic to an infrasolvmanifold;

(ii) $\chi(M) = 0$, π is torsion free and virtually poly-Z and $h(\pi) = 4$;

(iii) $\chi(M) = 0$ and π has a locally nilpotent normal subgroup ν such that $h(\nu) \geq 3$; and

(iv) $\chi(M) = 0$ *and* π *has an elementary amenable normal subgroup* ρ *with* $h(\rho) \geq 3$ *and whose finite subgroups have bounded order and which has no nontrivial finite normal subgroup, and* $H^2(\pi; Z[\pi]) = 0$.

Moreover, any torsion free virtually poly-Z group of Hirsch length 4 is the fundamental group of such a manifold, which is determined up to homeomorphism by the group.

Proof. If M is homeomorphic to an infrasolvmanifold then $\chi(M) = 0$ and π is torsion free and virtually poly-Z and $h(\pi) = 4$, by the remarks above. Thus *(i)* implies *(ii)*. Suppose that *(ii)* holds. Then $h(\sqrt{\pi}) \geq 3$, by Lemma 1, and π is a PD_4-group, by Theorems 9.9 and 9.10 of [Bi], so $H^s(\pi; Z[\pi]) = 0$ for $s < 4$. Therefore *(ii)* implies *(iii)* and *(iv)*.

Suppose that *(iii)* holds. The subset T of elements of finite order in ν is a characteristic subgroup of ν, by Proposition 5.2.7 of [Ro] and hence is normal in π. Moreover ν/T is a torsion free locally nilpotent group and $h(\nu/T) = h(\nu) \geq 3$. Therefore the group ring $Z[\pi/T]$ has a safe extension, by Theorem I.7. Also $H^s(\pi; Z[\pi]) = H^s(\pi/T; Z[\pi/T]) = 0$ for $s \leq 2$, by Theorem I.9. Therefore π/T is a PD_4-group over Q, by the Addendum to Theorem II.7, and so $3 \leq c.d._Q\nu/T \leq 4$ (as in any case $h(\nu/T) \leq c.d._Q\nu/T$). In particular, ν/T is locally nilpotent of class at most 3, and so is in fact nilpotent.

If $c.d._Q\nu/T = 3$ then $h(\nu/T) = c.d._Q\nu/T$ and there is a free $Q[\nu/T]$-module W such that $H^3(\nu/T; W) \neq 0$. Moreover ν/T has a finitely generated subgroup F of Hirsch length 3. Any such subgroup lies in a subnormal sequence $F \leq F\zeta(\nu/T) \leq \nu/T$. Two applications of the LHSSS show that if F has infinite index in ν/T then $H^s(\nu/T; W) = 0$ for $s \leq 3$. Therefore this index must be finite and so ν/T is finitely generated. Since it is torsion free and nilpotent of Hirsch length 3 it is a PD_3-group. The LHSSS for π/T as an extension of π/ν by ν/T (with coefficients $Q[\pi/T]$) then implies that $H^4(\pi/T; Q[\pi/T]) \cong H^1(\pi/\nu; H^3(\nu/T; Q[\pi/T])) \cong H^1(\pi/\nu; Q[\pi/\nu])$. Since $H^4(\pi/T; Q[\pi/T]) \cong Q$ it follows that π/ν has two ends. Thus π has a subgroup σ of finite index such that $\nu \leq \sigma$ and $\sigma/\nu \cong Z$. By Theorem A of [BS78] σ is an ascending HNN extension with base a finitely generated subgroup of ν. Therefore σ and hence π are constructable and so are virtually torsion free [BB76]. If $c.d.\nu/T = 4$ then $[\pi : \nu] = [\pi/T : \nu/T]$ is finite [St77], so ν is finitely generated, and therefore virtually poly-Z. Thus in either case π has a torsion free solvable subgroup of finite index and Hirsch length 4. An LHSSS argument then shows that $H^s(\pi; Z[\pi]) = 0$ for $s < 4$. In particular, (*iii*) implies (*iv*).

If (*iv*) holds then M is aspherical by Theorems I.7 and II.7, so π is a PD_4-group and $3 \leq h(\rho) \leq c.d.\rho \leq 4$. Therefore ρ is torsion free and hence is virtually solvable, by Theorem I.3. If $c.d.\rho = 4$ then $[\pi : \rho]$ is finite [St77] and so π is virtually solvable also. If $c.d.\rho = 3$ then $c.d.\rho = h(\rho)$ and so ρ is a duality group and is FP [Kr86]. Therefore $H^q(\rho; Q[\pi]) \cong H^q(\rho; Q[\rho]) \otimes Q[\pi/\rho]$ and is 0 unless $q = 3$. It then follows from the LHSSS for π as an extension of π/ρ by ρ (with coefficients $Q[\pi]$) that $H^4(\pi; Q[\pi]) \cong H^1(\pi/\rho; Q[\pi/\rho]) \otimes H^3(\rho; Q[\rho])$. Therefore $H^1(\pi/\rho; Q[\pi/\rho]) \cong Q$, so π/ρ has two ends and we again find that π is virtually solvable. In all cases π is torsion free and virtually poly-Z, by Theorem 9.23 of [Bi], and $h(\pi) = 4$. It is shown in [AJ76] that any such group is the fundamental group of an infrasolvmanifold. The uniqueness up to homeomorphism of manifolds with such a group and Euler characteristic 0 was proven in Theorem V.7. This shows that (*iv*) implies (*i*) and also proves the final assertion. //

Can the hypotheses on the fundamental group in (*iv*) be relaxed to "π has an elementary amenable normal subgroup ρ with $h(\rho) \geq 3$"? In particular, does the condition $h(\rho) \geq 3$ imply $H^2(\pi; Z[\pi]) = 0$? The examples $F \times S^1 \times S^1$ where $F = S^2$ or is a closed hyperbolic surface show that the condition that $h(\rho) > 2$ is necessary. (See also Theorems 11 - 13 below).

2. Distinguishing between the infrasolvmanifold geometries

Let M be an infrasolvmanifold with fundamental group π. The geometries of solvable Lie type are E^4, Nil^4, $Nil^3 \times E^1$, $Sol^4_{m,n}$, Sol^4_0 and Sol^4_1. Which of these geometries M admits is largely determined by the structure of $\sqrt{\pi}$. (See also Proposition 10.4 of [Wl86]). As a geometric structure on a manifold lifts to each covering space of the manifold it shall suffice to show that the geometries on suitable finite covering spaces (corresponding to subgroups of finite index in π) can be recognized.

We shall begin by describing the possible Hirsch-Plotkin radicals. For each natural number $q \geq 1$ let Γ_q be the group with presentation $< x, y, z | xz = zx, yz = zy, xy = z^q yx >$. Every such group Γ_q is torsion free and nilpotent of Hirsch length 3.

Theorem 3. *Let G be a finitely generated torsion free nilpotent group of Hirsch length $h(G) \leq 4$. Then either*
(i) G is free abelian; or
(ii) $h(G) = 3$ and $G \cong \Gamma_q$ for some $q \geq 1$; or
(iii) $h(G) = 4$, $\zeta G \cong Z^2$ and $G \cong \Gamma_q \times Z$ for some $q \geq 1$; or
(iv) $h(G) = 4$, $\zeta G \cong Z$ and $G/\zeta G \cong \Gamma_q$ for some $q \geq 1$.
In the latter case G has characteristic subgroups which are free abelian of rank 1, 2 and 3. In all cases G is an extension of Z by a free abelian normal subgroup.

Proof. The centre ζG is nontrivial and the quotient $G/\zeta G$ is again torsion free, by Proposition 5.2.19 of [Ro]. We may assume that G is not abelian, and hence that $G/\zeta G$ is not cyclic. Hence $h(G/\zeta G) \geq 2$, so $h(G) \geq 3$ and $1 \leq h(\zeta G) \leq h(G) - 2$.

If $h(G) = 3$ then $\zeta G \cong Z$ and $G/\zeta G \cong Z^2$. On choosing elements x and y representing a basis of $G/\zeta G$ and z generating ζG we quickly find that G is isomorphic to one of the groups Γ_q, and so is an extension of Z by Z^2.

If $h(G) = 4$ and $h(\zeta G) = 2$ then $\zeta G \cong Z^2$ and $G/\zeta G \cong Z^2$, so $G' \subseteq \zeta G$. Since G may be generated by elements x, y, t and u where x and y represent a basis of $G/\zeta G$ and t and u are central it follows easily that G' is infinite cyclic. Therefore ζG is not contained in G' and G has an infinite cyclic direct factor. Hence $G \cong Z \times \Gamma_q$, for some $q \geq 1$, and so G is an extension of Z by Z^3.

The remaining possibility is that $h(G) = 4$ and $h(\zeta G) = 1$. In this case $\zeta G \cong Z$ and $G/\zeta G$ is torsion free nilpotent of Hirsch length 3. If $G/\zeta G$ were abelian G' would also be infinite cyclic, and the pairing from $G/\zeta G \times G/\zeta G$ into G' defined by the commutator would be nondegenerate and skewsymmetric. But there are no such pairings on free abelian groups of odd rank. Therefore $G/\zeta G \cong \Gamma_q$, for some $q \geq 1$.

Let $\zeta_2 G$ be the preimage in G of $\zeta(G/\zeta G)$. Then $\zeta_2 G \cong Z^2$ and is a characteristic subgroup of G, so $C_G(\zeta_2 G)$ is also characteristic in G. The quotient $G/\zeta_2 G$ acts by conjugation on $\zeta_2 G$. Since $Aut(Z^2) = GL(2,Z)$ is virtually free and $G/\zeta_2 G \cong \Gamma_q/\zeta\Gamma_q \cong Z^2$ and since $\zeta_2 G \neq \zeta G$ it follows that $h(C_G(\zeta_2 G)) = 3$. Since $C_G(\zeta_2 G)$ is nilpotent and has centre of rank ≥ 2 it is abelian, and so $C_G(\zeta_2 G) \cong Z^3$. The preimage in G of the torsion subgroup of $G/C_G(\zeta_2 G)$ is torsion free, nilpotent of Hirsch length 3 and virtually abelian and hence is abelian. Therefore $G/C_G(\zeta_2 G) \cong Z$. //

Suppose first that $\sqrt{\pi} \cong Z^4$. Then $\sqrt{\pi}$ is the maximal abelian normal subgroup of π and $[\pi : \sqrt{\pi}]$ is finite, so M is a flat 4-manifold, i.e., has the geometry E^4. There are 75 such groups π. (See page 126 of [Wo]). All are solvable. (The action of π on $\sqrt{\pi}$ by conjugation induces an embedding of the holonomy group $H = \pi/\sqrt{\pi}$ in $GL(4,Z)$. The kernel of the reduction homomorphism from $SL(n,Z)$ to $SL(n,Z/pZ)$ is torsion free, for all $n \geq 1$ and all odd primes p. Since $SL(4,Z/pZ)$ has order $p^3(p^4-1)(p^4-p)(p^4-p^2)$ it follows easily that H has order dividing $2^9 3^2 5$. As any extension of $Z/5Z$ by Z^4 has nontrivial torsion the order of H must divide $2^9 3^2$. Hence H is solvable, by the Burnside p-q Theorem, Proposition 8.5.3 of [Ro]). However if G_6 is the fundamental group of the orientable flat 3-manifold with noncyclic holonomy then $\pi = G_6 \times Z$ is not poly-Z.

Suppose next that $\sqrt{\pi} \cong Z^3$. Then $h(\pi/\sqrt{\pi}) = 1$ and so π has a normal subgroup σ of finite index which is a semidirect product $\sqrt{\pi} \times_\theta Z$, where the action of a generator t of Z by conjugation on $\sqrt{\pi}$ is given by a matrix θ in $GL(3,Z)$. On replacing σ by a subgroup of index 2, if necessary, we may assume that θ has no negative eigenvalues and so is in $SL(3,Z)$. The characteristic polynomial of θ is $X^3 - mX^2 + nX - 1$, where $m = trace(\theta)$ and $n = trace(\theta^{-1})$. The matrix θ has infinite order, for otherwise the subgroup generated by $\sqrt{\pi}$ and a suitable power of t would be abelian of rank 4. If two of the eigenvalues are equal then they are all 1 and so $N = \theta - I$ is nilpotent, i.e., $N^3 = (\theta - I)^3 = 0$. As π is not virtually abelian $N \neq 0$. If $N^2 = 0$ then we may choose a basis v, w for $\ker N \cong Z^2$ so that $\mathrm{Im} N$ is generated by qw for some $q \geq 1$. It then follows easily that σ is isomorphic to $\Gamma_q \times Z$ and so is a discrete cocompact subgroup of the group $Nil^3 \times R$. Otherwise (if $N^2 \neq 0$) σ is a discrete cocompact subgroup of Nil^4 and is nilpotent. In both cases $\sqrt{\pi}$ is nonabelian and $h(\sqrt{\pi}) = 4$. (We shall consider these cases later).

Therefore we may suppose that the eigenvalues of θ are distinct. If they are all real (and hence positive) then π is a discrete cocompact subgroup of the group $Sol^4_{m,n}$. If θ has 1 as a simple eigenvalue the characteristic polynomial has the form $(X-1)(X^2-pX+1)$ and $m = n = p+1$, where $p = a + a^{-1}$ is the sum of the other eigenvalues. Since θ has infinite order and 1 is a simple eigenvalue $p > 2$, so $\lambda = X - 1$ and $\mu = X^2 - pX + 1$ are irreducible and relatively prime. If we consider $\sqrt{\pi}$ as a module over the ring $Z[X]$ in the obvious way we see that the submodule A generated by $\lambda\sqrt{\pi}$ and $\mu\sqrt{\pi}$ is a direct sum and has finite index in $\sqrt{\pi}$, since $\lambda\sqrt{\pi} \cong Z^2$, $\mu\sqrt{\pi} \cong Z$ and $\lambda\sqrt{\pi} \cap \mu\sqrt{\pi} = 0$. It follows that $A \times_{\theta|_A} Z$ is a subgroup of finite index in π which is isomorphic to a product $B \times Z$, and so the geometry is $Sol^3 \times E^1 = Sol^4_{m,m}$.

If two of the eigenvalues are not real then they are complex conjugates and π is a discrete cocompact subgroup of the group $Isom(Sol^4_0)$. (Note that they cannot be roots of unity since θ has infinite order, and therefore the real eigenvalue is not ± 1).

In each of these cases $h(\pi/\sqrt{\pi}) = 1$, so π is an extension of Z or D by a normal subgroup ν which contains $\sqrt{\pi} = Z^3$ as a subgroup of finite index. If $\pi/\sqrt{\pi}$ maps onto Z then M is the mapping torus of a self homeomorphism of a flat 3-manifold. Otherwise $\pi \cong A *_\nu B$ where $[A:\nu] = [B:\nu] = 2$ and so M is the union of two twisted I-bundles over flat 3-manifolds, and is doubly covered by such a mapping torus.

Suppose finally that $\sqrt{\pi}$ is nonabelian. If $h(\sqrt{\pi}) = 4$ then M is a $Nil^3 \times E^1$- or Nil^4-manifold. The cases can be distinguished by the rank of $\zeta\sqrt{\pi}$. (See Theorem 3). In each case $\sqrt{\pi}$ has a normal subgroup isomorphic to Z^3 and so M is finitely covered by the mapping torus of a self homeomorphism of a flat 3-manifold. If $h(\sqrt{\pi}) = 3$ then M is a Sol^4_1-manifold and is either the mapping torus of a self homeomorphism of a Nil^3-manifold or the union of two twisted I-bundles over Nil^3-manifolds.

3. Mapping tori of self homeomorphisms of E^3-manifolds

It follows from the above that a 4-dimensional infrasolvmanifold M admits one of the product geometries of type E^4, $Nil^3 \times E^1$ or $Sol^3 \times E^1$ if and only if $\pi_1(M)$ has a subgroup of finite index of the form $\nu \times Z$, where ν is abelian, nilpotent of class 2 or solvable but not virtually nilpotent, respectively. In the next two sections we shall examine when M is the mapping torus of a self homeomorphism of a 3-dimensional infrasolvmanifold. (Note that if M is orientable then it must be a mapping torus, by Lemma II.14 and Theorem V.11).

Theorem 4. *Let ν be the fundamental group of a flat 3-manifold, and let θ be an automorphism of ν. Then*
(i) $\sqrt{\nu}$ is the maximal abelian subgroup of ν and $\nu/\sqrt{\nu}$ embeds in $Aut(\sqrt{\nu})$;
(ii) $Out(\nu)$ is finite if and only if $[\nu : \sqrt{\nu}] > 2$;
(iii) the kernel of the restriction homomorphism from $Out(\nu)$ to $Aut(\sqrt{\nu})$ is finite;
(iv) if $[\nu : \sqrt{\nu}] = 2$ then $(\theta|_{\sqrt{\nu}})^2$ has 1 as an eigenvalue;
(v) if $[\nu : \sqrt{\nu}] = 2$ and $\theta|_{\sqrt{\nu}}$ has infinite order but all of its eigenvalues are roots of unity then $((\theta|_{\sqrt{\nu}})^2 - I)^2 = 0$.

Proof. It follows immediately from Theorem 3 that $\sqrt{\nu} \cong Z^3$ and is thus the maximal abelian subgroup of ν. The kernel of the homomorphism from ν to $Aut(\sqrt{\nu})$ determined by conjugation is the centralizer $C = C_\nu(\sqrt{\nu})$. As $\sqrt{\nu}$ is central in C and $[C : \sqrt{\nu}]$ is finite, C has finite commutator subgroup, by Schur's Theorem (Proposition 10.1.4 of [Ro]). Since C is torsion free it must be abelian and so $C = \sqrt{\nu}$. Hence $H = \nu/\sqrt{\nu}$ embeds in $Aut(\sqrt{\nu}) \cong GL(3, Z)$. (This is just the holonomy representation).

If H has order 2 then θ induces the identity on H; if H has order greater than 2 then some power of θ induces the identity on H, since $\sqrt{\nu}$ is a characteristic subgroup of finite index. The matrix $\theta|_{\sqrt{\nu}}$ then commutes with each element of the image of H in $GL(3, Z)$, and the remaining assertions follow from simple calculations, on considering the possibilities for π and H listed in Theorems 3.5.5 and 3.5.9 of [Wo]. //

Corollary. *The mapping torus $M(\phi) = N \times_\phi S^1$ of a self homeomorphism*

ϕ of a flat 3-manifold N is flat if and only if the outer automorphism $[\phi_*]$ induced by ϕ has finite order. //

If N is flat and $[\phi_*]$ has infinite order then $M(\phi)$ may admit one of the other product geometries $Sol^3 \times E^1$ or $Nil^3 \times E^1$; otherwise it must be a Nil^4-manifold. (The latter can only happen if $N = R^3/Z^3$, by part (v) of the theorem).

Theorem 5. *Let M be an infrasolvmanifold with fundamental group π such that $\sqrt{\pi} \cong Z^3$ and $\pi/\sqrt{\pi}$ is an extension of $D = (Z/2Z) * (Z/2Z)$ by a finite normal subgroup. Then M is a $Sol^3 \times E^1$-manifold.*

Proof. Let $p : \pi \to D$ be an epimorphism with kernel K containing $\sqrt{\pi}$ as a subgroup of finite index, and let t and u be elements of π whose images under p generate D and such that $p(t)$ generates an infinite cyclic subgroup of index 2 in D. Then there is an $N > 0$ such that the image of $s = t^N$ in $\pi/\sqrt{\pi}$ generates a normal subgroup. In particular, the subgroup generated by s and $\sqrt{\pi}$ is normal in π and gsg^{-1} and s^{-1} have the same image in $\pi/\sqrt{\pi}$. Let θ be the matrix of the action of s on $\sqrt{\pi}$, with respect to some basis $\sqrt{\pi} \cong Z^3$. Then θ is conjugate to its inverse, since gsg^{-1} and s^{-1} agree modulo $\sqrt{\pi}$. Hence one of the eigenvalues of θ is ± 1. Since π is not virtually nilpotent the eigenvalues of θ must be distinct, and so the geometry must be of type $Sol^3 \times E^1$. //

Corollary. *If M admits one of the geometries Sol_0^4 or $Sol_{m,n}^4$ with $m \neq n$ then it is the mapping torus of a self homeomorphism of R^3/Z^3. The group π is isomorphic to $Z^3 \times_\theta Z$ for some θ in $GL(3, Z)$ and is a metabelian poly-Z group.*

Proof. This follows immediately from Theorems 4 and 5. //

If $p : \pi \to D$ is an epimorphism with kernel K then $\pi \cong G*_K H$ where G and H are the preimages of the factors of D and $[G : K] = [H : K] = 2$. Conversely any such generalised free product of torsion free groups with amalgamation over index 2 subgroups isomorphic to K is a torsion free extension of D by K. We may use this observation to find examples of E^4-, Nil^4-, $Nil^3 \times E^1$- and

$Sol^3 \times E^1$-manifolds which are not mapping tori. For instance, the groups with presentations

$$< u, v, x, y, z | xy = yx, xz = zx, yz = zy, uxu^{-1} = x^{-1}, u^2 = y, uzu^{-1} = z^{-1},$$

$$v^2 = z, vxv^{-1} = x^{-1}, vyv^{-1} = y^{-1} >,$$

$$< u, v, x, y, z | xy = yx, xz = zx, yz = zy, u^2 = x, uyu^{-1} = y^{-1}, uzu^{-1} = z^{-1},$$

$$v^2 = x, vyv^{-1} = v^{-4}y^{-1}, vzv^{-1} = z^{-1} >,$$

and

$$< u, v, x, y, z | xy = yx, xz = zx, yz = zy, u^2 = x, v^2 = y, uyu^{-1} = x^4y^{-1},$$

$$vxv^{-1} = x^{-1}y^2, uzu^{-1} = vzv^{-1} = z^{-1} >$$

are each generalised free products of two copies of $Z^2 \times_{-I} Z$ amalgamated over their maximal abelian subgroups. The Hirsch-Plotkin radicals of these groups are isomorphic to Z^4 (generated by $\{(uv)^2, x, y, z\}$), $\Gamma_2 \times Z$ (generated by $\{uv, x, y, z\}$) and Z^3 (generated by $\{x, y, z\}$), respectively. The group with presentation

$$< u, v, x, y, z | xy = yx, xz = zx, yz = zy, u^2 = x, uz = zu, uyu^{-1} = x^2y^{-1},$$

$$v^2 = y, vxv^{-1} = x^{-1}, vzv^{-1} = v^4z^{-1} >$$

is a generalised free product of copies of $(Z \times_{-1} Z) \times Z$ (generated by $\{u, y, z\}$) and $Z^2 \times_{-I} Z$ (generated by $\{v, x, z, \}$) amalgamated over their maximal abelian subgroups. Its Hirsch-Plotkin radical is the subgroup of index 4 generated by $\{(uv)^2, x, y, z\}$, and is nilpotent of class 3. The manifolds corresponding to these groups admit the geometries E^4, $Nil^3 \times E^1$, $Sol^3 \times E^1$ and Nil^4, respectively. However they cannot be mapping tori, as these groups each have finite abelianization.

4. Mapping tori of self homeomorphisms of Nil^3-manifolds

The arguments in this section are analogous to those of Section 3. In particular, the next theorem is largely parallel to Theorem 4.

Theorem 6. *Let ν be the fundamental group of a Nil^3-manifold N. Then*
(i) $\nu/\sqrt{\nu}$ embeds in $Aut(\sqrt{\nu}/\zeta\sqrt{\nu}) \cong GL(2,Z)$;
(ii) $\bar{\nu} = \nu/\zeta\sqrt{\nu}$ is a 2-dimensional crystallographic group;
(iii) $Out(\bar{\nu})$ is infinite if and only if $\bar{\nu} \cong Z^2$ or $Z^2 \times_{-I} (Z/2Z)$;
(iv) the kernel of the natural homomorphism from $Out(\nu)$ to $Out(\bar{\nu})$ is finite.

Proof. Let C be the kernel of the homomorphism from ν to $Aut(\sqrt{\nu}/\zeta\sqrt{\nu})$ determined by conjugation. Then $\sqrt{\nu}/\zeta\sqrt{\nu}$ is central in $C/\zeta\sqrt{\nu}$ and $[C/\zeta\sqrt{\nu} : \sqrt{\nu}/\zeta\sqrt{\nu}]$ is finite, so $C/\zeta\sqrt{\nu}$ has finite commutator subgroup, by Schur's Theorem (Proposition 10.1.4 of [Ro]). Since C is torsion free it follows easily that C is nilpotent and hence that $C = \sqrt{\nu}$. This proves (*i*) and (*ii*), and (*iii*) now follows as in Theorem 4, on considering the possible finite subgroups of $GL(2,Z)$. (See page 85 of [Z]).

If $\zeta\nu \neq 1$ then $\zeta\nu = \zeta\sqrt{\nu} \cong Z$ and so the kernel of the natural homomorphism from $Aut(\nu)$ to $Aut(\bar{\nu})$ is isomorphic to $Hom(\nu/\nu', Z)$. If ν/ν' is finite this kernel is trivial; otherwise $\bar{\nu} \cong Z^2$, $Z \times_{-1} Z$, $Z \times D$ or $D \times_\tau Z$ (where τ is the automorphism of $D = (Z/2Z) * (Z/2Z)$ which interchanges the factors). If $\bar{\nu} \cong Z^2$ then $\nu = \sqrt{\nu} \cong \Gamma_q$, for some $q \geq 1$, and so the kernel is isomorphic to $(Z/qZ)^2$. (See page 101 of [H]). In the remaining cases $H^2(\bar{\nu}; Z)$ is finite and so any central extension of such a group by Z is virtually abelian. Thus (*iv*) holds for such groups.

If $\zeta\nu = 1$ then $\nu/\sqrt{\nu} < GL(2,Z)$ has an element of order 2 with determinant -1. This element cannot be conjugate to $\left(\begin{smallmatrix} 0 & 1 \\ 1 & 0 \end{smallmatrix}\right)$, for otherwise ν would not be torsion free. On considering the list on page 85 of [Z] we see that the image of $\nu/\sqrt{\nu}$ in $GL(2,Z)$ is conjugate to a subgroup of the group of diagonal matrices $\left(\begin{smallmatrix} \epsilon & 0 \\ & \epsilon' \end{smallmatrix}\right)$, with $|\epsilon| = |\epsilon'| = 1$. If $\nu/\sqrt{\nu}$ is generated by $\left(\begin{smallmatrix} 1 & 0 \\ 0 & -1 \end{smallmatrix}\right)$ then $\nu/\zeta\sqrt{\nu} \cong Z \times_{-1} Z$ and $\nu \cong Z^2 \times_\theta Z$, where $\theta = \left(\begin{smallmatrix} -1 & r \\ 0 & -1 \end{smallmatrix}\right)$ for some nonzero integer r, and N is a circle bundle over the Klein bottle. If $\nu/\sqrt{\nu} \cong (Z/2Z)^2$ then ν has a presentation $< t, u, z | u^2 = z, tzt^{-1} = z^{-1}, ut^2u^{-1} = t^{-2}z^s >$, and N is a Seifert bundle over the orbifold $P(22)$. It may be verified in each case that the kernel of the natural homomorphism from $Out(\nu)$ to $Out(\bar{\nu})$ is finite. Therefore (*iv*) is true for all Nil^3-manifold groups. //

In fact every Nil^3-manifold is a Seifert bundle over a 2-dimensional euclidean orbifold [Sc83'].

Theorem 7. *The mapping torus $M(\phi) = N \times_\phi S^1$ of a self homeomorphism ϕ of a Nil^3-manifold N is orientable, and is a $Nil^3 \times E^1$-manifold if and only if the outer automorphism $[\phi_*]$ induced by ϕ has finite order.*

Proof. Since N is orientable and ϕ is orientation preserving (see pages 101-102 of [H]), the mapping torus $M(\phi)$ must be orientable.

The subgroup $\zeta\sqrt{\nu}$ is characteristic in ν and hence normal in π, and $\nu/\zeta\sqrt{\nu}$ is virtually Z^2. If $M(\phi)$ is a $Nil^3 \times E^1$-manifold then $\pi/\zeta\sqrt{\nu}$ is also virtually abelian. It follows easily that that the image of ϕ_* in $Aut(\nu/\zeta\sqrt{\nu})$ has finite order. Hence $[\phi_*]$ has finite order also, by Theorem 6. Conversely, if $[\phi_*]$ has finite order in $Out(\nu)$ then π has a subgroup of finite index which is isomorphic to $\nu \times Z$, and so $M(\phi)$ has the product geometry, by the discussion above. //

Theorem 4.2 of [KLR83] (which extends Bieberbach's theorem to the virtually nilpotent case) may be used to show directly that every outer automorphism class of finite order of the fundamental group of an E^3- or Nil^3-manifold is realizable by an isometry of an affinely equivalent manifold.

The image of an automorphism θ of Γ_q in $Out(\Gamma_q)$ has finite order if and only if the induced automorphism $\bar{\theta}$ of $\bar{\Gamma}_q = \Gamma_q/\zeta\Gamma_q \cong Z^2$ has finite order in $Aut(\bar{\Gamma}_q) \cong GL(2,Z)$. If $\bar{\theta}$ has infinite order but has trace ± 2 (i.e., if $\bar{\theta}^2 - I$ is a nonzero nilpotent matrix) then $\pi = \Gamma_q \times_\theta Z$ is virtually nilpotent of class 3. If the trace of $\bar{\theta}$ has absolute value greater than 2 then $h(\sqrt{\pi}) = 3$.

Theorem 8. *Let M be a closed 4-manifold which admits one of the geometries Nil^4-manifold M or Sol^4_1. Then M is the mapping torus of a self homeomorphism of a Nil^3-manifold if and only if it is orientable.*

Proof. If M is such a mapping torus then it is orientable, by Theorem 7. Conversely, if M is orientable then $\pi = \pi_1(M)$ has infinite abelianization, by Lemma II.14. Let $p : \pi \to Z$ be an epimorphism with kernel K, and let t be an element of π such that $p(t)$ generates Z. If K is virtually nilpotent of class 2 we are done, by Theorem V.11. (Note that this must be the case if M is a Sol^4_1-manifold). If K is virtually abelian then $K \cong Z^3$, by part (v) of Theorem 4. The matrix corresponding to the action of t on K by conjugation must be orientation preserving, since M is orientable. It follows easily that π is nilpotent. Hence there is another epimorphism with kernel nilpotent of class 2, and so the theorem is proven. //

Let π be a torsion free virtually nilpotent group of Hirsch length 4. Then $\sqrt{\pi}$ has a characteristic subgroup C which is free abelian of rank 2. Since C is characteristic in $\sqrt{\pi}$ it is normal in π. Let $\tilde{F}$ be the preimage in π of the maximal finite subgroup of π/C. Then $\tilde{F}$ is torsion free and $h(\tilde{F}) = h(C) = 2$ so $\tilde{F}$ is isomorphic to Z^2 or $Z \times_{-1} Z$, and $\pi/\tilde{F}$ is a 2-dimensional crystallographic group. It follows easily that $[\pi : \sqrt{\pi}] \leq 24$. If $\sqrt{\pi}$ is nilpotent of class 3 we can do better, for then π has a subgroup of index 2 which is a semidirect product $Z^3 \times_\theta Z$. (This uses part (v) of Theorem 4). Since $(\theta^2 - I)$ is a nilpotent matrix it follows that $[\pi : \sqrt{\pi}] \leq 4$.

The 4-manifold corresponding to $\sqrt{\pi}$ is both the mapping torus of a self homeomorphism of R^3/Z^3 and also the mapping torus of a self homeomorphism of a Nil^3-manifold. We have already seen that $Nil^3 \times E^1$- and Nil^4-manifolds need not be mapping tori at all. We shall round out this discussion with examples illustrating the remaining combinations of mapping torus structure and orientation compatible with Lemma II.14 and Theorem 8 above. As the groups have abelianization of rank 1 the corresponding manifolds are mapping tori in an essentially unique way. The groups with presentations

$$< x, y, z, t \mid xy = yx, xz = zx, yz = zy, txt^{-1} = x^{-1}, tyt^{-1} = y^{-1},$$

$$tzt^{-1} = yz^{-1} >,$$

$$< x, y, z, t \mid z = [x, y], xz = zx, yz = zy, txt^{-1} = x^{-1}, tyt^{-1} = y^{-1} >$$

and

$$< x, y, z, t \mid xy = yx, zxz^{-1} = x^{-1}, zyz^{-1} = y^{-1}, txt^{-1} = x^{-1},$$

$$ty = yt, tzt^{-1} = z^{-1} >$$

are each virtually nilpotent of class 2. The corresponding $Nil^3 \times E^1$-manifolds are mapping tori of self homeomorphisms of R^3/Z^3, a flat 3-manifold and a

Nil^3-manifold, respectively. The latter two of these manifolds are orientable. The groups with presentations

$$< x, y, z, t | xy = yx, xz = zx, yz = zy, txt^{-1} = x^{-1}, tyt^{-1} = xy^{-1},$$

$$tzt^{-1} = yz^{-1} >$$

and

$$< x, y, z, t | z = [x, y], xz = zx, yz = zy, txt^{-1} = x^{-1}, tyt^{-1} = xy^{-1} >$$

are each virtually nilpotent of class 3. The corresponding Nil^4-manifolds are mapping tori of self homeomorphisms of R^3/Z^3 and of a Nil^3-manifold, respectively.

The group with presentation

$$< t, u, x, y, z | [x, y] = z^2, xz = zx, yz = zy, txt^{-1} = x^2y, tyt^{-1} = xy, tz = zt,$$

$$u^4 = z, uxu^{-1} = y^{-1}, uyu^{-1} = x, utu^{-1} = t^{-1} >$$

has Hirsch-Plotkin radical isomorphic to Γ_2 (generated by $\{x, y, z\}$), and has finite abelianization. The corresponding Sol_1^4-manifold is nonorientable and is not a mapping torus.

5. Mapping tori of self homeomorphisms of Sol^3-manifolds

The arguments in this section are again analogous to those of Section 3.

Theorem 9. *Let σ be the fundamental group of a Sol^3-manifold. Then*
(i) $\sqrt{\sigma} \cong Z^2$ and $\sigma/\sqrt{\sigma} \cong Z$ or D;
(ii) $Out(\sigma)$ is finite.

Proof. The argument of Lemma 1 implies that $h(\sqrt{\sigma}) > 1$. Since σ is not virtually nilpotent $h(\sqrt{\sigma}) < 3$. Hence $\sqrt{\sigma} \cong Z^2$, by Theorem 3. Let $\tilde{F}$ be the preimage in σ of the maximal finite normal subgroup of $\sigma/\sqrt{\nu}$, let t be an element of σ whose image generates the maximal abelian subgroup of $\sigma/\tilde{F}$ and let τ be the automorphism of $\tilde{F}$ determined by conjugation by t. Let σ_1 be the subgroup of σ generated by $\tilde{F}$ and t. Then $\sigma_1 \cong \tilde{F} \times_\tau Z$, $[\sigma : \sigma_1] \leq 2$,

$\tilde{F}$ is torsion free and $h(\tilde{F}) = 2$. If $\tilde{F} \neq \sqrt{\sigma}$ then $\tilde{F} \cong Z \times_{-1} Z$. But extensions of Z by $Z \times_{-1} Z$ are virtually abelian, since $Out(Z \times_{-1} Z)$ is finite. Hence $\tilde{F} = \sqrt{\sigma}$ and so $\sigma/\sqrt{\sigma} \cong Z$ or D.

Every automorphism of σ induces automorphisms of $\sqrt{\sigma}$ and of $\sigma/\sqrt{\sigma}$. Let $Out^+(\sigma)$ be the subgroup of $Out(\sigma)$ represented by automorphisms which induce the identity on $\sigma/\sqrt{\sigma}$. The restriction of any such automorphism to $\sqrt{\sigma}$ commutes with τ. We may view $\sqrt{\sigma}$ as a module over the ring $R = Z[X]/(\lambda(X))$, where $\lambda(X) = X^2 - tr(\tau)X + det(\tau)$ is the characteristic polynomial of τ. The polynomial λ is irreducible and has real roots which are not roots of unity, for otherwise $\sqrt{\sigma} \times_\tau Z$ would be virtually nilpotent. Therefore R is a domain and its field of fractions $Q[X]/(\lambda(X))$ is a real quadratic number field. The R-module $\sqrt{\sigma}$ is clearly finitely generated, R-torsion free and of rank 1. Hence the endomorphism ring $End_R(\sqrt{\sigma})$ is a subring of $\tilde{R}$, the integral closure of R. Since $\tilde{R}$ is the ring of integers in $Q[X]/(\lambda(X))$ the group of units $\tilde{R}^\times$ is isomorphic to $\{\pm 1\} \times Z$. Since τ determines a unit of infinite order in $R^\times$ the index $[\tilde{R}^\times : \tau^Z]$ is finite.

Suppose now that $\sigma/\sqrt{\sigma} \cong Z$. If f is an automorphism which induces the identity on $\sqrt{\sigma}$ and on $\sigma/\sqrt{\sigma}$ then $f(t) = tw$ for some w in $\sqrt{\sigma}$. If w is in the image of $\tau - 1$ then f is an inner automorphism. Now $\sqrt{\sigma}/(\tau-1)\sqrt{\sigma}$ is finite, of order $det(\tau - 1)$. Since τ is the image of an inner automorphism of σ it follows that $Out^+(\sigma)$ is an extension of a subgroup of $\tilde{R}^\times/\tau^Z$ by $\sqrt{\sigma}/(\tau-1)\sqrt{\sigma}$. Hence $Out(\sigma)$ has order dividing $2[\tilde{R}^\times : \tau^Z]det(\tau - 1)$.

If $\sigma/\sqrt{\sigma} \cong D$ then σ has a characteristic subgroup σ_1 such that $[\sigma : \sigma_1] = 2$, $\sqrt{\sigma} < \sigma_1$ and $\sigma_1/\sqrt{\sigma} \cong Z = \sqrt{D}$. Every automorphism of σ restricts to an automorphism of σ_1. It is easily verified that the restriction from $Aut(\sigma)$ to $Aut(\sigma_1)$ is a monomorphism. Since $Out(\sigma_1)$ is finite it follows that $Out(\sigma)$ is also finite. //

Corollary. *The mapping torus of a self homeomorphism of a Sol^3-manifold is a $Sol^3 \times E^1$-manifold.* //

The group σ with presentation $< x, y, t | xy = yx, txt^{-1} = x^3y^2, tyt^{-1} = x^2y >$ is the fundamental group of a nonorientable Sol^3-manifold Σ. The nonorientable $Sol^3 \times E^1$-manifold $\Sigma \times S^1$ is the mapping torus of id_Σ and is also the mapping torus of a self homeomorphism of R^3/Z^3. The groups with presentations

$$< x, y, z, t | xy = yx, zxz^{-1} = x^{-1}, zyz^{-1} = y^{-1}, txt^{-1} = xy, tyt^{-1} = x,$$
$$tzt^{-1} = z^{-1} >,$$
$$< x, y, z, t | xy = yx, zxz^{-1} = x^2y, zyz^{-1} = xy, tx = xt, tyt^{-1} = x^{-1}y^{-1},$$
$$tzt^{-1} = z^{-1} >,$$
$$< x, y, z, t | xy = yx, xz = zx, yz = zy, txt^{-1} = x^2y, tyt^{-1} = xy, tzt^{-1} = z^{-1} >$$

and

$$< x, y, t, u | xy = yx, txt^{-1} = x^2y, tyt^{-1} = xy, uxu^{-1} = y^{-1}, uyu^{-1} = x,$$
$$utu^{-1} = t^{-1} >$$

have Hirsch-Plotkin radical Z^3 and abelianization of rank 1. The corresponding $Sol^3 \times E^1$-manifolds are mapping tori in an essentially unique way. The first two are orientable, and are mapping tori of self homeomorphisms of the orientable flat 3-manifold with holonomy of order 2 and of an orientable Sol^3-manifold, respectively. The latter two are nonorientable, and are mapping tori of orientation reversing self homeomorphisms of R^3/Z^3 and of the same orientable Sol^3-manifold, respectively.

6. Other aspherical product geometries

For the remaining geometries (apart from the spherical-euclidean geometries and those with compact models) it is not yet known whether the s-cobordism theorem holds, nor whether $L_5(\pi, w)$ acts on the structure set, for the fundamental groups arising all contain nonabelian free subgroups. Motivated by the group theoretic terminology we shall say that a manifold M *virtually* has some property if it has a finite covering space which has that property.

In the next theorem we use 3-manifold theory to extend the characterization of aspherical mapping tori with product geometries to the cases $H^3 \times E^1$,

$\tilde{SL} \times E^1$ and $H^2 \times E^2$. (We shall consider the remaining aspherical product geometry $H^2 \times H^2$ in the next section, with the other semisimple geometries).

Theorem 10. *Let ϕ be a self homeomorphism of a 3-manifold N with a geometry of type $H^2 \times E^1$ or $\tilde{SL}$. Then the mapping torus $M(\phi) = N \times_\phi S^1$ admits the corresponding product geometry if and only if the outer automorphism $[\phi_*]$ induced by ϕ has finite order. The mapping torus of a self homeomorphism ϕ of a hyperbolic 3-manifold N which is also Haken admits the geometry $H^3 \times E^1$.*

Proof. Let $\nu = \pi_1(N)$ and let t be an element of $\pi = \pi_1(M(\phi))$ which projects to a generator of $\pi_1(S^1)$. If $M(\phi)$ has geometry $\tilde{SL} \times E^1$ then π is a discrete cocompact subgroup of $Isom(\tilde{SL}) \times Isom(E^1)$. On passing to the 2-fold covering space $M(\phi^2)$, if necessary, we may assume that π is a discrete cocompact subgroup of $Isom(\tilde{SL}) \times R$. The radical of this group is R^2 and is central, and π meets the radical in a lattice subgroup L, by Proposition 8.27 of [Rg]. Since the centre of ν is Z the image of L in π/ν is nontrivial. It follows easily that π has a subgroup σ of finite index which is isomorphic to $\nu \times Z$, and in particular that conjugation by $t^{[\pi:\sigma]}$ induces an inner automorphism of ν.

If $M(\phi)$ has geometry $H^2 \times E^2$ then π is a discrete cocompact subgroup of $Isom(H^2) \times Isom(E^2)$. In this case π meets $Isom(E^2)$ in a (torsion free) discrete cocompact subgroup and thus has a subgroup σ of finite index which is isomorphic to $\rho \times Z^2$, where ρ is a discrete cocompact subgroup of $Isom(H^2)$, and is a subgroup of ν. (See Section 2 of [Wl86]). It again follows that $t^{[\pi:\sigma]}$ induces an inner automorphism of ν.

Conversely, suppose that N has a geometry of type $H^2 \times E^1$ or $\tilde{SL}$ and that $[\phi_*]$ has finite order in $Out(\nu)$. Then ϕ is homotopic to a self homeomorphism of (perhaps larger) finite order [Zn80] and is therefore isotopic to such a self homeomorphism [Sc85,BO91], which may be assumed to preserve the geometric structure [MS86]. Thus we may assume that ϕ is an isometry. The self homeomorphism of $N \times R$ sending (n, r) to $(\phi(n), r+1)$ is then an isometry for the product geometry and the mapping torus has the product geometry.

If N is hyperbolic then by Mostow rigidity ϕ is homotopic to an isometry of finite order [Mo68], and if also N is Haken then ϕ is isotopic to such an isometry [Wd68], and again the mapping torus has the product geometry. //

If $[\phi_*]$ has infinite order and N is an $H^2 \times E^1$-manifold then $M(\phi)$ may admit the geometry $H^3 \times E^1$. However if $[\phi_*]$ has infinite order and N is an $\tilde{SL}$-manifold then $M(\phi)$ admits no geometric structure. If $\zeta\nu \cong Z$ and $\zeta(\nu/\zeta\nu) = 1$ then $Hom(\nu/\nu', \zeta\nu)$ embeds in $Out(\nu)$, and thus ν has outer automorphisms of infinite order, in most cases [CR77].

Theorem 11. *Let M be a closed 4-manifold with $\chi(M) = 0$ and fundamental group π, and suppose that π has a torsion free elementary amenable normal subgroup ρ with $h(\rho) = 2$ and which has infinite index in π. Then M is aspherical and ρ is virtually abelian.*

Proof. Since ρ is torsion free elementary amenable and $h(\rho) = 2$ it is virtually solvable, by Theorem I.3. Therefore $A = \sqrt{\rho}$ is nontrivial, and as it is characteristic in ρ it is normal in π. Since A is torsion free and $h(A) \leq 2$ it is abelian, by Theorem 3.

Suppose first that $h(A) = 1$. Then A is isomorphic to a subgroup of Q and the homomorphism from $B = \rho/A$ to $Aut(A)$ induced by conjugation in ρ is injective. Since $Aut(A)$ is isomorphic to a subgroup of $Q^\times$ and $h(B) = 1$ either $B \cong Z$ or $B \cong Z \oplus (Z/2Z)$. We must in fact have $B \cong Z$, since A is the maximal abelian subgroup of ρ and is torsion free. Moreover A is not finitely generated and the centre of ρ is trivial. The quotient group π/A has one end as the image of ρ is an infinite cyclic normal subgroup of infinite index. Therefore π is 1-connected at ∞, by Theorem 1 of [Mi87], and so $H^s(\pi; Z[\pi]) = 0$ for $s \leq 2$ [GM86]. Hence M is aspherical and π is a PD_4-group, by the Corollary to Theorem II.7.

As A is a characteristic subgroup every automorphism of ρ restricts to an automorphism of A. This restriction from $Aut(\rho)$ to $Aut(A)$ is an epimorphism, with kernel isomorphic to A, and so $Aut(\rho)$ is solvable. Let $C = C_\pi(\rho)$ be the centralizer of ρ in π. Then C is nontrivial, for otherwise π would be isomorphic to a subgroup of $Aut(\rho)$ and hence would be virtually poly-Z. But then A would be finitely generated, ρ would be virtually abelian and $h(A) = 2$. Moreover $C \cap \rho = \zeta\rho = 1$, so $C\rho \cong C \times \rho$ and $c.d.C + c.d.\rho = c.d.C\rho \leq c.d.\pi = 4$. The quotient group $\pi/C\rho$ is isomorphic to a subgroup of $Out(\rho)$.

If $c.d.C\rho \leq 3$ then as C is nontrivial and $h(\rho) = 2$ we must have $c.d.C = 1$ and $c.d.\rho = h(\rho) = 2$. Therefore C is free and ρ is of type FP [Kr86]. By Theorem A of [BS78] ρ is an ascending HNN group with base a finitely generated subgroup of A and so has a presentation of the form $< a, t|tat^{-1} = a^n >$ for some nonzero integer n. We may assume $n > 1$, as ρ is not virtually abelian. The subgroup of $Aut(\rho)$ represented by $(n-1)A$ consists of inner automorphisms. Since $n > 1$ the quotient $A/(n-1)A \cong Z/(n-1)Z$ is finite, and as $Aut(A) \cong Z[1/n]^\times$ it follows that $Out(\rho)$ is virtually abelian. Therefore π has a subgroup σ of finite index which contains $C\rho$ and such that $\sigma/C\rho$ is a finitely generated free abelian group, and in particular $c.d.\sigma/C\rho$ is finite. As σ is a PD_4-group it follows from Theorem 9.11 of [Bi] that $C\rho$ is a PD_3-group and hence that ρ is a PD_2-group. We reach the same conclusion if $c.d.C\rho = 4$, for then $[\pi : C\rho]$ is finite [St77] and so $C\rho$ is a PD_4-group. As a solvable PD_2-group is virtually abelian our original assumption must have been wrong.

Therefore $h(A) = 2$ and so every finitely generated subgroup of ρ of Hirsch length 2 is isomorphic to either Z^2 or $Z \times_{-1} Z$. Hence $[\rho : A] \leq 2$. If A is finitely generated then $H^q(A; Z[\pi]) = 0$ for $q < 2$ and $H^2(A; Z[\pi]) \cong Z[\pi/A]$, so $H^s(\pi; Z[\pi]) = 0$ for $s \leq 2$, by an LHSSS argument. If A is not finitely generated then it is an extension of an infinite torsion group by a free abelian subgroup of rank 2, so another LHSSS argument gives $H^q(A; Z[\pi]) = 0$ for $q \leq 2$ and so again $H^s(\pi; Z[\pi]) = 0$ for $s \leq 2$. Since $Z[\pi]$ has a safe extension, by Theorem I.7, M is aspherical, by the Corollary to Theorem II.7. //

Every group with a presentation of the form $< a, t|tat^{-1} = a^n >$ is torsion free and solvable, and is the fundamental group of some orientable closed 4-manifold M with $\chi(M) = 0$. Thus the hypothesis that the subgroup ρ have infinite index in π is necessary for the above theorem.

Theorem 12. *Let M, π and ρ be as in Theorem 11. Then π has a normal subgroup N which is free abelian of rank 2 and $H^2(\pi/N; Z[\pi/N]) \cong Z$. If*

ρ is not finitely generated then π/N is a finitely presentable, infinite torsion group and π has a subgroup of finite index isomorphic to $Z \times \sigma$, where σ is a PD_3-group whose centre is not finitely generated.

Proof. Since M is aspherical π is a PD_4-group. If π has such a normal subgroup $N \cong Z^2$ the corresponding LHSSS with coefficients $Z[\pi]$ has only one nonzero row and thus gives an isomorphism $H^2(\pi/N; Z[\pi/N]) \cong H^4(\pi; Z[\pi]) \cong Z$.

Let $A = \sqrt{\rho}$. Then A is abelian of rank 2 and $[\rho : A]$ is finite, so $A \cong Z^2$ if and only if ρ is finitely generated, and we may then take $N = A$. In particular, this is the case if π is elementary amenable, for then it is virtually poly-Z, by Theorem 2, and so each of its subgroups is finitely generated. Therefore we may assume without loss of generality that $h(\sqrt{\pi}) = 2$ and that $A = \rho = \sqrt{\pi}$ and is not finitely generated.

Suppose first that $[\pi : C] = \infty$, where $C = C_\pi(A)$. Then $c.d.C \leq 3$ [St77]. Since A is not finitely generated $c.d.A = h(A) + 1 = 3$, by Theorem 7.14 of [Bi]. Hence $C = A$, by Theorem 8.8 of [Bi], so the homomorphism from π/A to $Aut(A)$ determined by conjugation in π is a monomorphism. Since A is torsion free abelian of rank 2 $Aut(A)$ is isomorphic to a subgroup of $GL(2, Q)$ and therefore any torsion subgroup of $Aut(A)$ is locally finite. (See page 105 of [K]). Thus if $\pi'A/A$ is a torsion group then $\pi'A$ is elementary amenable and so π is itself elementary amenable, contradicting our assumption. Hence we may suppose that there is an element g in π' which has infinite order modulo A. The subgroup $< A, g >$ generated by A and g is an extension of Z by A and has infinite index in π, for otherwise π would be virtually solvable. Hence $c.d. < A, g >= 3 = h(< A, g >)$ [St77]. By Theorem 7.15 of [Bi], $L = H_2(A; Z)$ is the underlying abelian group of a subring $Z[m^{-1}]$ of Q, and the action of g on L is multiplication by a rational number a/b, where a and b are relatively prime and ab and m have the same prime divisors. But g acts on A as an element of $GL(2, Q)' \leq SL(2, Q)$. Since $L = A \wedge A$, by Proposition 11.4.16 of [Ro], g acts on L via $det(g) = 1$. Therefore $m = 1$ and so L must be finitely generated. But then A must also be finitely generated, again contradicting our assumption.

Thus we may assume that C has finite index in π. Let $A_1 < A$ be a subgroup of A which is free abelian of rank 2. Then A_1 is central in C and C/A_1 is finitely presentable. Since $[\pi : C]$ is finite A_1 has only finitely many distinct conjugates in π, and they are all subgroups of ζC. Let N be their product. Then N is a finitely generated torsion free abelian normal subgroup of π and $2 \leq h(N) \leq h(\sqrt{C}) \leq h(\sqrt{\pi}) = 2$.

Suppose that there is an element g in π whose image in π/N has infinite order. Then the image of $h = g^{[\pi:C]}$ in π/A also has infinite order, since A/N is a torsion group. The subgroup $< A, h >$ generated by A and h is isomorphic to $A \times Z$, and so $c.d. < A, h >= 4$. But then $< A, h >$ has finite index in π [St77] and so A is finitely generated, contradicting our hypothesis. Therefore π/N is a finitely presentable torsion group, and it is infinite since $c.d.\pi = 4 > c.d.N = 2$.

Let M^+ be the orientable covering space with fundamental group $\pi^+ = \ker w_1(M)$. Since $\chi(M^+) = 0$ there is an epimorphism $\lambda : \pi^+ \to Z$, by Lemma II.14. Let $\sigma = \ker\lambda$ and let $A^+ = A \cap \pi^+$. If $A^+ \leq \sigma$ then $A^+ = C_\sigma(A^+)$, by Theorem 8.8 of [Bi], since $3 = c.d.A^+ \leq c.d.C_\sigma(A^+) \leq c.d.\sigma < 4$. But then π would be virtually solvable and hence virtually poly-Z, which would contradict our assumption. Thus $A^+\sigma$ has finite index in π, and so is a PD_4-group. As it is clearly isomorphic to $Z \times \sigma$ and as $A^+ \cap \sigma$ is central in σ the theorem is proven. //

The arguments of Chapter V of [H] may be adapted to give further partial information on the structure of the fundamental groups of manifolds as in Theorems 11 and 12. The evidence seems to suggest that a closed 4-manifold M is virtually s-cobordant to the total space of a torus bundle over a surface if and only if $\chi(M) = 0$, π has a torsion free abelian normal subgroup of rank 2 and also has a subgroup of finite index which has abelianization of rank at least 2. There are no known examples of finitely presentable, infinite torsion groups.

It has recently been shown that if the fundamental group of an irreducible closed orientable 3-manifold has infinite centre then the centre is finitely generated [Me88] and the 3-manifold is Seifert fibred [Ga91]. The corresponding result for PD_3-groups holds under the additional hypothesis that there be a subgroup of finite index with infinite abelianization. (See [Hi85] and Theorem 6 of the Appendix). This hypothesis would be redundant if it could be shown that $H^2(G;Z[G]) \cong Z$ implies that G is virtually a PD_2-group. As a consequence we may characterize $H^2 \times E^2$- and $\tilde{SL} \times E^1$-manifolds up to virtual simple homotopy type in terms of their fundamental groups.

Theorem 13. *Let M be a closed 4-manifold with fundamental group π. Then M is virtually simple homotopy equivalent to an $H^2 \times E^2$- or $\tilde{SL} \times E^1$-manifold if and only if $\chi(M) = 0$, π has a torsion free abelian normal subgroup A of rank 2 such that π/A is not virtually abelian and π has a normal subgroup of finite index isomorphic to $Z \times \sigma$ where σ/σ' is infinite. The geometry is of type $H^2 \times E^2$ if and only if there is such a subgroup ρ which is itself isomorphic to a product $Z \times \tau$.*

Proof. The conditions are necessary, for a manifold with such a geometry is finitely covered by the cartesian product of a surface of hyperbolic type with a torus or by the product of an $\tilde{SL}$-manifold with a circle. (Compare the first paragraph of Theorem 10).

Suppose that they hold. On replacing π by a subgroup of finite index, if necessary, we may assume that M is orientable and $\pi = Z \times \sigma$ where $\sigma \cap A = Z$ and is central in σ, $\sigma/\sigma \cap A$ is infinite and σ/σ' is infinite. Since M is aspherical, by Theorem 11, π is a PD_4-group and so σ is a PD_3-group. The other assumptions on σ then imply that it is the fundamental group of an aspherical Seifert fibred 3-manifold N, which must be of type $H^2 \times E^1$ or $\tilde{SL}$ since $\sigma/\sigma \cap A$ is not virtually abelian. ([Hi85] - see also Theorem 6 of the Appendix). Note that N is also Haken, since σ/σ' is infinite. Since M is aspherical there is a homotopy equivalence $f : M \to N \times S^1$, which is simple as $Wh(\pi) = 0$, by Lemma 1.1 of [St84]. The final assertion is clear. //

Ue has shown that orientable $H^2 \times E^2$- and $\tilde{SL} \times E^1$-manifolds are determined up to diffeomorphism (among such geometric manifolds) by their fundamental groups [Ue91].

An argument similar to that of Theorem 13 shows that a 4-manifold M with fundamental group π is virtually simple homotopy equivalent to an $H^3 \times E^1$-manifold if and only if $\chi(M) = 0$, $\sqrt{\pi} = Z$ and π has a normal subgroup of finite index which is isomorphic to $Z \times \rho$ where ρ is a discrete cocompact subgroup of $PSL(2, C)$. If every PD_3-group is the fundamental group of an aspherical closed 3-manifold and if every atoroidal aspherical closed 3-manifold is hyperbolic we could replace the last assertion by the more intrinsic conditions that ρ have one end (which would suffice with the other conditions to imply that M is aspherical and hence that ρ is a PD_3-group) and no noncyclic abelian subgroups (which would imply that any irreducible 3-manifold with fundamental group ρ is atoroidal).

7. Semisimple geometries

There are three other aspherical geometries which are realizable by closed 4-manifolds, namely $H^2 \times H^2$, H^4 and $H^2(C)$. (The model spaces for $H^2 \times H^2$ and $H^2(C)$ may be taken as the unit polydisc $\{(w, z) \in C^2 : |w| < 1, |z| < 1\}$ and unit ball $\{(w, z) \in C^2 : |w|^2 + |z|^2 < 1\}$, respectively). Every closed manifold admitting one of these geometries has positive Euler characteristic [Ko92]. If M is an $H^2 \times H^2$- or H^4-manifold then $\sigma(M) = 0$, while if M is an $H^2(C)$-manifold it is orientable and $\chi(M) = 3\sigma(M) > 0$ [Wl86].

If a manifold M is simple homotopy equivalent to one with one of these geometries then $\pi = \pi_1(M)$ must be isomorphic to a discrete torsion free cocompact subgroup of the isometry group and $\chi(M) = \chi(\pi)$. These conditions are also sufficient if the classifying map $c_M : M \to K(\pi, 1) = \pi\backslash X$ has nonzero degree, by Theorem II.5, which implies that c_M is then a homotopy equivalence, together with the work of Farrell and Jones [FJ90], which shows that $Wh(\pi) = 0$ and that the surgery obstruction maps are isomorphisms. However, at present there are not even conjectural intrinsic characterizations of the fundamental groups arising in these cases. We have only the following partial result.

Theorem 14. *Let M be a closed 4-manifold with fundamental group π. Then M is virtually simple homotopy equivalent to the cartesian product of two H^2-manifolds if and only if π is virtually a PD_4-group, $\sqrt{\pi} = 1$ and π has a torsion free normal subgroup of finite index which is isomorphic to a nontrivial product $\sigma \times \tau$ where $\chi(M)[\pi : \sigma \times \tau] = (2 - \beta_1(\sigma))(2 - \beta_1(\tau))$.*

Proof. The conditions are clearly necessary. Suppose that they hold. On replacing π by a normal subgroup of finite index, if necessary, we may assume that M is orientable, $\pi \cong \sigma \times \tau$ and π is a PD_4-group of orientable type. Since $\sqrt{\pi} = 1$ neither factor can be infinite cyclic, and so σ and τ are each PD_2-groups. Since $\sqrt{\sigma} = \sqrt{\tau} = 1$, these groups are fundamental groups of closed orientable H^2-manifolds, F_σ and F_τ, say. It now follows from Lemma V.2 that M is simple homotopy equivalent to $F_\sigma \times F_\tau$, which is clearly an $H^2 \times H^2$-manifold. //

Let $P = PSL(2, R)$ and let Γ be a discrete cocompact subgroup of $P \times P$. Then $M = \Gamma \backslash H^2 \times H^2$ is a compact complex analytic surface. If $\Gamma \cap P \times \{1\}$ or $\Gamma \cap \{1\} \times P$ is nontrivial then we may argue as in Theorem 6.3 of [Wl85] to conclude that M is virtually a cartesian product. Otherwise the natural foliations of $H^2 \times H^2$ descend to give a pair of transverse foliations of M by copies of H^2. (Conversely, if M is a closed Riemannian 4-manifold with a codimension 2 metric foliation by totally geodesic surfaces then M has a finite cover which either admits the geometry $H^2 \times E^2$ or $H^2 \times H^2$ or is the total space of an S^2 or $S^1 \times S^1$-bundle over a closed surface or is the mapping torus of a self homeomorphism of the 3-torus R^3/Z^3, $S^2 \times S^1$ or a lens space [Ca89]).

The group of units of a quaternion algebra over a real quadratic field F which splits at both infinite primes may be viewed as a discrete cocompact subgroup of $PSL(2, F_{\infty_1}) \times PSL(2, F_{\infty_2}) \cong P \times P$, and in [Wl86] it is observed that in such a case M is not virtually a cartesian product. Thus not every $H^2 \times H^2$-manifold satisfies the hypotheses of Theorem 14. (See also page 177 of [BPV]).

CHAPTER VII

MANIFOLDS COVERED BY $S^2 \times R^2$

In Chapter II we saw that if the universal covering space of a closed 4-manifold with infinite fundamental group is homotopy equivalent to a finite complex then it is either contractible or homotopy equivalent to S^2 or S^3. The cases when M is aspherical and π is elementary amenable were dealt with in Chapter VI. Here we shall show that if $\tilde{M} \simeq S^2$ then $\tilde{M}$ is homeomorphic to $S^2 \times R^2$ and π is the group of a 4-manifold with geometry $S^2 \times E^2$ or $S^2 \times H^2$, provided that π satisfies a hypothesis which may well be redundant. We shall then examine in detail the case when π is elementary amenable (corresponding to the geometry $S^2 \times E^2$). There are nine such groups, and each is realized by only finitely many homotopy types of manifolds covered by $S^2 \times R^2$. Six of these groups have infinite abelianization, and for these the homotopy types may be distinguished by their Stiefel-Whitney classes.

1. Virtually geometric manifolds

In this section we shall show that if the fundamental group of a manifold with universal covering space $S^2 \times R^2$ has a subgroup of finite index with infinite abelianization then it may be realized by a closed geometric 4-manifold.

Theorem 1. *Let M be a closed 4-manifold such that $\pi = \pi_1(M)$ has a subgroup of finite index with infinite abelianization and $\pi_2(M) \cong Z$. Let $u : \pi \to \{\pm 1\} = Aut(\pi_2(M))$ be the natural homomorphism. Then*

(i) the covering space corresponding to $\kappa = \mathrm{Ker}u \cap \mathrm{Ker}w_1(M)$ is s-cobordant to the total space of an S^2-bundle over an aspherical closed orientable surface and the universal covering space is homeomorphic to $S^2 \times R^2$;

(ii) π is the fundamental group of a closed manifold admitting the geometry $S^2 \times E^2$, if π is virtually Z^2, or $S^2 \times H^2$ otherwise.

Typeset by $\mathcal{A}_{\mathcal{M}}\mathcal{S}$-TEX

Proof. The first assertion follows from Theorems II.11, IV.3 and V.13. If π is torsion free then it is itself a surface group. If π has a nontrivial finite normal subgroup then it is a direct product $\kappa \times (Z/2Z)$. In either case π is the fundamental group of a corresponding product of surfaces. Otherwise π is a semidirect product $\kappa \tilde{\times} (Z/2Z)$ and is a plane motion group, by a theorem of Nielsen [Ni42]. (See also Theorem A of [EM82]). This means that there is a monomorphism $f : \pi \to Isom(X)$ with image a discrete subgroup which acts cocompactly on X, where X is the Euclidean or hyperbolic plane, according as π is virtually abelian or not. The homomorphism $(u, f) : \pi \to \{\pm I\} \times Isom(X) \leq Isom(S^2 \times X)$ is then a monomorphism onto a discrete subgroup which acts freely and cocompactly on $S^2 \times R^2$. In all cases such a group may be realised geometrically. //

The orbit space of the geometric action of π described above is a cartesian product with S^2 if u is trivial and fibres over RP^2 otherwise.

Lemma 2. *Let ξ be an S^2-bundle over a closed surface F and let $q : F' \to F$ be a 2-fold covering map with connected domain F'. Then $w_2(q^*\xi) = 0$.*
Proof. It is sufficient to show that $w_2(q^*\xi) \cap [F'] = 0$, by Poincaré duality for F'. As q_* is an isomorphism in degree 0 and q has degree 2, this follows from the projection formula, which gives $q_*(w_2(q^*\xi) \cap [F']) = q_*((q^* w_2(\xi)) \cap [F']) = w_2(\xi) \cap q_*[F'] = 0$. //

This lemma also follows from the more geometric approach of [Me84].

Theorem 3. *Let M be a closed 4-manifold with fundamental group π. Then M is virtually s-cobordant to an $S^2 \times H^2$-manifold if and only if π is virtually a PD_2-group, has no infinite abelian normal subgroup and $\chi(M) = 2\chi(\pi)$. If these conditions hold M has a covering space of degree dividing 4 which is s-cobordant to a product $S^2 \times B$, where B is a closed orientable hyperbolic surface.*

Proof. Let Γ be a discrete cocompact subgroup of $Isom(S^2 \times H^2)$. On passing to a subgroup of finite index, if necessary, we may assume that Γ is a subgroup of $SO(3) \times PSL(2, R)$. Since $\Gamma \cap SO(3) \times \{1\}$ acts freely and by orientation preserving maps on $S^2 \times H^2$ it is trivial. Therefore projection onto the second factor maps Γ monomorphically to a discrete cocompact subgroup of $PSL(2,R)$, and so it has a torsion free subgroup γ of finite index. (See Theorem 4.10.1 of [ZVC]). The quotient space $\gamma\backslash S^2 \times H^2$ is the total space of an S^2-bundle over a closed orientable surface of hyperbolic type. Thus the conditions are necessary.

Suppose that they hold. On passing to a 2-fold covering space, if necessary, we may assume that M is s-cobordant to the total space of an S^2-bundle ξ with base an aspherical closed surface, by Theorem 1. On again passing to a 2-fold covering space and applying Lemma 2, if necessary, we may assume that $w_2(\xi) = w_1(\xi) = 0$. Hence ξ is trivial and M is s-cobordant to a product $S^2 \times B$, where B is a closed orientable surface. Since π has no nontrivial abelian normal subgroup B is of hyperbolic type. //

The 5-dimensional s-cobordism theorem has been established so far only over elementary amenable fundamental groups [FQ]. In the present context much weaker assumptions on the group imply that it is in fact virtually abelian.

Theorem 4. *Let M be a closed 4-manifold with fundamental group π. Then the following conditions are equivalent:*
(i) π is virtually Z^2 and $\chi(M) = 0$;
(ii) π has an infinite amenable normal subgroup and $\pi_2(M) \cong Z$;
(iii) $\chi(M) = 0$ and $\pi_2(M) \cong Z$; and
(iv) M is virtually homeomorphic to an $S^2 \times E^2$-manifold.
If these conditions hold M has a covering space of degree dividing 4 which is homeomorphic to $S^2 \times S^1 \times S^1$.

Proof. Note that these conditions are invariant under passage to finite covers and subgroups of finite index in π. If (i) holds then the covering space of M with fundamental group Z^2 is homeomorphic to the total space of an S^2-bundle over the torus and so $\pi_2(M) \cong Z$, by Theorems IV.3 and V.10. As in Theorem 3 above, after passing to a finite covering, if necessary, we may assume this bundle is trivial and $M \cong S^2 \times S^1 \times S^1$. Thus (i) implies each of (ii), (iii), and (iv).

Suppose that π has an infinite amenable normal subgroup A and $\pi_2(M) \cong Z$. Then $\tilde{M} \simeq S^2$ and we may assume that π is torsion free and $H^2(\pi; Z[\pi]) \cong Z$, by Theorem II.11. If A has finite index in π then π is itself amenable and so $\chi(M) = 0$, by Theorem 1.2 of [Ec92]. If $[\pi : A] = \infty$ then A is free, by Theorem 2.2 of [Fa74], and so $A \cong Z$. Therefore the group ring $Z[\pi]$ has a safe extension R, by Theorem I.7. Thus $R \otimes_{Z[\pi]} H_*(\tilde{M}; Z) = 0$ and so a localization argument shows that $\chi(M) = 0$. Hence *(ii)* implies *(iii)*.

Suppose that *(iii)* holds. On passing to a finite covering if necessary we may assume that M is orientable and that π acts trivially on $\pi_2(M)$. Hence π/π' is infinite, by Lemma II.14, and M is s-cobordant to the total space of an S^2-bundle over an aspherical surface, by Theorem 1. Since $\chi(M) = 0$ the surface must be the torus or the Klein bottle, and so π is virtually Z^2. Hence *(iii)* implies *(i)* and so *(iv)*.

An orientable manifold is an $S^2 \times E^2$-manifold if and only if it is homeomorphic to the mapping torus of an orientation preserving self homeomorphism of $S^2 \times S^1$ or $RP^3 \sharp RP^3$, by Theorem 12 of [Ue91]. Since any such mapping torus has $S^2 \times S^1 \times S^1$ as a 4-fold covering space it is clear that *(iv)* implies each of the other conditions.

The final assertion follows as in Theorem 3. //

Corollary. *Any two of the conditions $\chi(M) = 0$, π is virtually Z^2 and $\pi_2(M) \cong Z$ imply the third.//*

We may use a localization argument instead of appealing to [Ec92], if the subgroup A is elementary amenable. Can the hypothesis on π in condition *(ii)* be relaxed to "π has an infinite normal subgroup which has no noncyclic free subgroup"? (This suffices if also the subgroup has infinite index, as is clear from the proof).

2. Fundamental groups of $S^2 \times E^2$-manifolds

We shall assume henceforth that the conditions of Theorem 4 hold, and in the next theorem we shall show that there are nine possible groups. Seven of them are 2-dimensional euclidean orbifold groups, and we shall give also the name of the corresponding orbifold, following Appendix A of [M]. (Note that the restriction on finite subgroups eliminates the remaining ten of the seventeen orbifold groups from consideration).

Theorem 5. *Let M be a closed 4-manifold such that $\pi = \pi_1(M)$ is virtually Z^2 and $\chi(M) = 0$. Let A and F be the maximal abelian and maximal finite normal subgroups (respectively) of π. If π is torsion free then either*
(i) $\pi = A \cong Z^2$ (the torus), or
(ii) $\pi \cong Z \times_{-1} Z$ (the Klein bottle).
If $F = 1$ but π has nontrivial torsion and $[\pi : A] = 2$ then either
(iii) $\pi \cong D \times Z \cong (Z \oplus (Z/2Z)) *_Z (Z \oplus (Z/2Z))$, with the presentation
$< u, v, x | u^2 = v^2 = 1, ux = xu, vx = xv >$ *(the silvered annulus), or*
(iv) $\pi \cong D \times_\tau Z \cong Z *_Z (Z \oplus (Z/2Z))$, with the presentation
$< t, u | t^2 u = ut^2, u^2 = 1 >$ *(the silvered Möbius band), or*
(v) $\pi \cong D *_Z D$, with the presentation
$< s, u, w | u^2 = w^2 = 1, usu = wsw = s^{-1} >$ *(the pillowcase $S(2222)$).*
If $F = 1$ and $[\pi : A] = 4$ then either
(vi) $\pi \cong D *_Z (Z \oplus (Z/2Z))$, with the presentation
$< s, t, u | t^2 = u^2 = 1, usu = s^{-1}, ts = st >$ *($D(22)$), or*
(vii) $\pi \cong Z *_Z D$, with the presentation $< t, u | u^2 = 1, ut^2 u = t^{-2} >$ ($P(22)$).
If F is nontrivial then either
(viii) $\pi \cong Z^2 \oplus (Z/2Z)$, or
(ix) $\pi \cong (Z \times_{-1} Z) \times (Z/2Z)$.

Proof. Let $u : \pi \to \{\pm 1\} = Aut(\pi_2(M))$ be the natural homomorphism. Since Ker$u$ is torsion free it is either Z^2 or $Z \times_{-1} Z$; since it has index at most 2 it follows that $[\pi : A]$ divides 4 and that F has order at most 2. If $F = 1$ then $A \cong Z^2$ and π/A acts effectively on A, so π is a 2-dimensional crystallographic group. If $F \neq 1$ then it is central in π and u maps F isomorphically to $Z/2Z$, so $\pi \cong (Z/2Z) \times$Ker$u$. //

We may use the combinatorial structure of these groups to show that they all have trivial Whitehead group.

Lemma 6. *Let π be a 2-dimensional crystallographic group which has no finite subgroup of order greater than 2. Then $Wh(\pi) = 0$.*

Proof. If $G = A *_C B$ (or $A*_C$) is a generalized free product with amalgamation (or HNN extension) in which $Wh(A) = Wh(B) = Wh(C) = 0$ and the integral group ring $Z[C]$ is a regular coherent ring then $Wh(G) = 0$, by the Mayer-Vietoris sequence of Waldhausen [Wd78]. Since $D = (Z/2Z) * (Z/2Z)$ and $Wh(Z/2Z) = 0$ this implies that $Wh(D) = 0$; since $Wh(Z \oplus (Z/2Z)) = Wh(Z) = 0$ [Kw86] and $Z[Z]$ is a regular noetherian ring and each of these groups is such a generalized free product the lemma follows. //

We do not know whether the other 2-dimensional crystallographic groups have trivial Whitehead groups; they cannot be expressed as free products with amalgamation over Z. (Note however that the finite subgroups of these groups all have trivial Whitehead group).

If Γ is a discrete subgroup of $Isom(S^2 \times E^2) = O(3) \times E(2) = O(3) \times (R^2 \tilde{\times} O(2))$ which acts freely and cocompactly on $S^2 \times R^2$ then $F = \Gamma \cap (O(3) \times \{1\})$ is its maximal finite normal subgroup and its image $p_2(\Gamma)$ under the projection $p_2 : Isom(S^2 \times E^2) \to E(2)$ is a 2-dimensional crystallographic group. Therefore $p_2(\Gamma)$ is determined up to conjugacy in $Aff(2)$ by the isomorphism type of the group Γ/F.

Let $(A, \beta, C) \in O(3) \times (R^2 \tilde{\times} O(2))$ be the isometry which sends $(v, X) \in S^2 \times R^2$ to $(Av, CX + \beta)$.

3. The homotopy type

Our next result shows that if M satisfies the conditions of Theorem 4 and its fundamental group has infinite abelianization then it is determined up to homotopy by $\pi_1(M)$ and its Stiefel-Whitney classes.

Theorem 7. *Let M be a closed 4-manifold with $\chi(M) = 0$ and such that $\pi = \pi_1(M)$ is virtually Z^2. If π/π' is infinite then M is homotopy equivalent to an $S^2 \times E^2$-manifold which fibres over S^1.*

Proof. The infinite cyclic covering space of M determined by an epimorphism $\lambda : \pi \to Z$ is a PD_3-complex, by Theorem III.4, and therefore is homotopy equivalent to $S^2 \times S^1$ (if $\mathrm{Ker}\lambda \cong Z$ is torsion free and $w_1(M)|_{\mathrm{Ker}\lambda} = 0$), $S^2 \tilde{\times} S^1$ (if $\mathrm{Ker}\lambda \cong Z$ and $w_1(M)|_{\mathrm{Ker}\lambda} \neq 0$), $RP^2 \times S^1$ (if $\mathrm{Ker}\lambda \cong Z \oplus (Z/2Z)$)

or $RP^3\sharp RP^3$ (if $\mathrm{Ker}\lambda \cong D$). Therefore M is homotopy equivalent to the mapping torus $M(\phi)$ of a self homotopy equivalence of one of these spaces.

The group of free homotopy classes of self homotopy equivalences $E(S^2 \times S^1)$ is generated by the reflections in each factor and the twist map, and has order 8. The group $E(S^2\tilde{\times}S^1)$ has order 4 [KR90]. Two of the corresponding mapping tori also arise from self homeomorphisms of $S^2 \times S^1$. The other two have nonintegral w_1. These mapping tori (with π torsion free) are also S^2-bundles over the torus or Klein bottle.

Let $R_i \in O(3)$ be the reflection of R^3 which changes the sign of the ith coordinate, for $i = 1,2,3$. If A and B are products of such reflections then the subgroups of $Isom(S^2 \times E^2)$ generated by $(A,(1,0),I)$ and $(B,(0,1),I)$ are discrete, isomorphic to Z^2 and act freely and cocompactly on $S^2 \times R^2$. Taking (*i*) $A = B = I$, (*ii*) $A = I, B = -I$, (*iii*) $A = R_1R_2, B = R_1R_3$ and (*iv*) $A = R_1, B = R_1R_2$ gives the four S^2-bundles over the torus. If instead we use the isometries $(A,(1,0),\left(\begin{smallmatrix}1 & 0\\ 0 & -1\end{smallmatrix}\right))$ and $(B,(0,1),I)$ we obtain discrete subgroups isomorphic to $Z\tilde{\times}Z$ which act freely and cocompactly. Taking (*v*) $A = B = I$, (*vi*) $A = -I, B = I$, (*vii*) $A = I, B = -I$, (*viii*) $A = R_1R_2, B = R_1R_3$, (*ix*) $A = R_1, B = R_1R_2$ and (x) $A = R_1R_2, B = R_1$ gives the six S^2-bundles over the Klein bottle.

The group $E(RP^2 \times S^1)$ is generated by the reflection in the second factor and by a twist map, and has order 4. The mapping tori are also RP^2-bundles over the torus or Klein bottle. Adjoining the fixed point free involution $(-I,0,I)$ to any one of the above ten sets of generators for the S^2-bundle groups amounts to dividing out the S^2 fibres by the antipodal map and so we obtain the corresponding RP^2-bundles. (Note that there are just four such RP^2-bundles - but each has several distinct double covers which are S^2-bundles).

The group $E(RP^3\sharp RP^3)$ is generated by the reflection interchanging the summands and the fixed point free involution (cf. page 251 of [Ba']), and has order 4. Let $\alpha = (-I,0,\left(\begin{smallmatrix}-1 & 0\\ 0 & 1\end{smallmatrix}\right))$, $\beta = (I,(1,0),I)$, $\gamma = (I,(0,1),I)$ and $\delta = (-I,(0,1),I)$. Then the subgroups generated by $\{\alpha,\beta,\gamma\}$, $\{\alpha,\beta,\delta\}$, $\{\alpha,\beta\gamma\}$ and $\{\alpha,\beta\delta\}$, respectively, give the four $RP^3\sharp RP^3$-bundles.

In the S^2- and RP^2-bundle cases each group and orientation character occurs twice in our list. The corresponding manifolds may be distinguished by whether or not $w_2(M) = 0$. (See Section 4 of [Ue91], where this is done for the orientable cases (*i*) and (*iii*), and (*v*) and (*viii*), respectively). As all the mapping tori are geometrically realizable, the theorem is proven. //

The S^2- and RP^2-bundle cases could also be approached from the point of view of Chapter IV.

When π is torsion free every homomorphism from π to $Z/2Z$ arises as the orientation character for some M with fundamental group π. However if $\pi \cong D \times Z$ or $D \times_\tau Z$ the orientation character must be trivial on the subgroup D while if $F \neq 1$ the orientation character is uniquely determined. In all cases, to each choice of orientation character there corresponds a unique action of π on $\pi_2(M)$. However the homomorphism from π to $Z/2Z$ determining the action may differ from $w_1(M)$.

The possible orientation characters for the groups with finite abelianization are restricted by Lemma II.14, which implies that the kernel of $w_1 : \pi \to Z/2Z$ must have infinite abelianization. For $D *_Z D$ we must have $w_1(u) = w_1(w) = 1$ and $w_1(s) = 0$. For $D *_Z (Z \oplus (Z/2Z))$ we must have $w_1(s) = 0$ and $w_1(u) = 1$; since the subgroup generated by the commutator subgroup and t is isomorphic to $D \times Z$ we must also have $w_1(t) = 0$. Thus the orientation characters are uniquely determined for these groups. For $Z *_Z D$ we must have $w_1(u) = 1$, but $w_1(t)$ may be either 0 or 1. As there is an automorphism ϕ of $Z *_Z D$ determined by $\phi(t) = ut$ and $\phi(u) = u$ we may assume that $w_1(t) = 0$ in this case.

We shall show next that each of these possibilities may be realised geometrically.

Lemma 8. *Let $J = (A, \beta, C) \in O(3) \times E(2)$ be an isometry of order 2 which is fixed point free. Then $A = -I$. If moreover J is orientation reversing then $C = I$ and $\beta = 0$ or $C = -I$.*

Proof. Since $J^2 = (A^2, C\beta + \beta, C^2)$ we must have $C^2 = I$ and $C\beta + \beta = 0$ and so the involution (β, C) has the fixed point $\beta/2 \in R^2$. Therefore $A \in O(3)$ must be a fixed point free involution, and so $A = -I$. If J is orientation reversing then $det(C) = 1$; as $C^2 = I \in O(2)$ this implies that either $C = I$ (and hence $\beta = 0$) or $C = -I$. //

The isometries $S = (I, (1,0), I)$, $U = (-I, 0, -I)$ and $W = (-I, (0,1), -I)$ generate a discrete subgroup of $Isom(S^2 \times E^2)$ isomorphic to $D *_Z D$ and which acts freely and cocompactly on $S^2 \times R^2$. The isometries $S = (I, (1,0), I)$, $T = (-I, 0, \left(\begin{smallmatrix} 1 & 0 \\ 0 & -1 \end{smallmatrix}\right))$ and $U = (-I, (0,1), -I)$ generate such a subgroup isomorphic to $D *_Z (Z \oplus (Z/2Z))$. The isometries $T = (\tau, (1,0), \left(\begin{smallmatrix} 1 & 0 \\ 0 & -1 \end{smallmatrix}\right))$ and $U = (-I, (0,1), -I)$ generate such a subgroup isomorphic to $Z *_Z D$, provided that $\tau^2 = I$ (so that T^2U is a fixed point free involution). Up to conjugation there are four possible values for τ, and two of these (namely $\tau = -I$ and $\tau = R_1$) determine orientation preserving isometries T. Moreover, by Lemma 8 and the remarks following Lemma 6 if M is an $S^2 \times E^2$-manifold and π/π' is finite then π is conjugate in $O(3) \times Aff(2)$ to one of these subgroups, and so M is diffeomorphic to the corresponding quotient space.

If π/π' is finite no manifold with such a fundamental group can fibre over S^1. However if Γ is a discrete subgroup of $Isom(S^2 \times E^2)$ which acts freely on $S^2 \times R^2$ and is generated by isometries of the form $(\pm I, \beta, C)$ then the map which sends $(v, X) \in S^2 \times R^2$ to $[\pm v] \in RP^2$ is compatible with the action of Γ and so the $S^2 \times E^2$-manifold $S^2 \times R^2/\Gamma$ fibres over RP^2. If at least one of the generators is the free involution on the S^2-factor then the fibre is $R^2/\bar{\Gamma}$, where $\bar{\Gamma} = \Gamma \cap (\{1\} \times E(2))$. (Otherwise the fibre has two components). This is clearly the case for three of the above manifolds.

An $S^1 \times S^1$-bundle over RP^2 which does not fibre over S^1 has fundamental group $D *_Z D$, while a K-bundle over RP^2 which does not fibre over S^1 has group $D *_Z (Z \oplus (Z/2Z))$ or $Z *_Z D$ (assuming throughout that π is virtually Z^2). In general, must a closed 4-manifold M with $\chi(M) = 0$ and such a fundamental group be homotopy equivalent to the total space of a bundle over RP^2? Is every such bundle space geometric?

We may adapt the arguments of [HK88] to obtain a somewhat weaker result for the remaining cases.

Theorem 9. *Let M be a closed 4-manifold with $\chi(M) = 0$ and whose fundamental group π is virtually Z^2. Then the homotopy type of M is determined* *up to a finite ambiguity by* π.

Proof. If π/π' is infinite the result follows from Theorem 8, so we may suppose that π/π' is finite and hence that M is nonorientable. As $f_M : M \to P_2(M)$ is 3-connected we may define a class w in $H^1(P_2(M); Z/2Z)$ by $f_M^* w = w_1(M)$. Let $S_4^{PD}(P_2(M))$ be the set of "polarized" PD_4-complexes (X, f) where $f : X \to P_2(M)$ is 3-connected and $w_1(X) = f^* w$, modulo homotopy equivalence over $P_2(M)$. (Note that $\chi(X) = 0$ and so the universal cover of X is homotopy equivalent to S^2). Let Z^w be the local system determined by w (considered as a homomorphism from π to $Aut(Z)$), and let $[X]$ be the fundamental class of X in $H_4(X; Z^w)$. It follows as in Lemma 1.3 of [HK88] that given two such polarized complexes (X, f) and (Y, g) there is a map $h : X \to Y$ with $gh = f$ if and only if $f_*[X] = g_*[Y]$ in $H_4(P_2(M); Z^w)$. Since $\tilde{X} \simeq \tilde{Y} \simeq S^2$ and f and g are 3-connected such a map h must be a homotopy equivalence.

The action of π on $\pi_2(M)$ is unique if $\pi \cong D *_Z D$ or $D *_Z (Z \oplus (Z/2Z))$, since these groups are generated by involutions (which must act nontrivially). As the subgroup of $Z *_Z D$ generated by its commutator subgroup and t is isomorphic to $Z \tilde{\times} Z$ the action of t on $\pi_2(M)$ is determined once $w_1(t)$ is known. As $\pi \cong \rho \tilde{\times} (Z/2Z)$ where $c.d.\rho = 2$ a calculation with the LHSSS shows that $H^3(\pi; \pi_2(M))$ is finite (of order dividing 2^4). A comparison of the spectral sequences of the universal covering maps of X and $P_2(M)$ shows that $H_4(P_2(M); Z^w) \cong H_4(X; Z^w) \oplus H_4(P_2(M), X; Z^w) \cong H_4(X; Z^w) \oplus (H_0(\pi; H_4(K(Z,2); Z) \otimes Z^w))$; the first summand is infinite cyclic, and is generated by $f_*[X]$, while as $w \neq 0$ the second summand is of order 2. Hence there are only finitely many values for $f_*[X]$. Together these observations prove the theorem. //

Is the homotopy type of M completely determined by $\pi_1(M)$, $w(M)$ and $k_1(M)$? Note that (in each case) the k-invariant restricts to the k-invariant of $\tilde{M}/ < u >$, which is the nonzero element of $H^3(Z/2Z; \pi_2(M))$. Thus there are at most eight possible values for the k-invariant.

4. Some remarks on the homeomorphism types

In Chapter V we showed that if π is torsion free then M must be homeomorphic to the total space of an S^2-bundle over the torus or Klein bottle, and we were able to estimate the size of the structure sets when π has $Z/2Z$ as a direct factor. In the remaining cases the groups are not "square-root closed accessible" and we have not been able to compute the surgery obstruction groups completely. However the Mayer-Vietoris sequences of [Ca73] are exact modulo 2-torsion, and we may compare the ranks of $[SM; G/TOP]$ and $L_5(\pi, w_1)$. This is sufficient in some cases to show that the structure set is infinite. For instance, the rank of $L_5(D\times Z)$ is 3 and that of $L_5(D\times_\tau Z)$ is 2, while the rank of $L_5(D*_Z(Z\oplus(Z/2Z)), w_1)$ is 2. If M is orientable and $\pi \cong D\times Z$ or $D\times_\tau Z$ then $[SM; G/TOP] \cong H^3(M;Z) \oplus H^1(M;Z/2Z) \cong H_1(M;Z) \oplus H^1(M;Z/2Z)$ has rank 1. Therefore $S_{TOP}(M)$ is infinite. If $\pi \cong D *_Z (Z \oplus (Z/2Z))$ then $H_1(M;Q) = 0$, $H_2(M;Q) = H_2(\pi;Q) = 0$ and $H_4(M;Q) = 0$, since M is nonorientable. Hence $H^3(M;Q) \cong Q$, since $\chi(M) = 0$. Therefore $[SM; G/TOP]$ again has rank 1 and $S_{TOP}(M)$ is infinite. These estimates do not suffice to decide whether there are infinitely many homeomorphism classes in the homotopy type of M. To decide this we need to study the action of the group $E(M)$ on $S_{TOP}(M)$. A scheme for analyzing $E(M)$ as a tower of extensions involving actions of cohomology groups with coefficients determined by Whitehead's Γ-functors is outlined on page 52 of [Ba'].

CHAPTER VIII

MANIFOLDS COVERED BY $S^3 \times R$

In this chapter we shall show that if the universal covering space of a closed 4-manifold is homotopy equivalent to S^3 then it is homeomorphic to $S^3 \times R$, and the manifold is finitely covered by $S^3 \times S^1$. The fundamental group is an extension of Z or D by a finite normal subgroup F. It follows easily from work of Hambleton and Madsen that F must be either the group of an S^3-manifold or one of the groups $Q(8a,b,c) \times Z/dZ$ with a,b,c and d odd. (There are examples of the latter type, and no such M is homotopy equivalent to an $S^3 \times E^1$-manifold). We are not completely successful in determining the possible fundamental groups, even when the subgroup F is an S^3-manifold group. The homotopy type is determined by the fundamental group and a k-invariant; we do not know which k-invariants are realizable. Finally, computing the Whitehead groups and surgery obstruction groups is a major task and probably the best we can hope for are estimates in the style of page 137 of [H].

1. Covering spaces

In Theorem II.10 we saw that the universal covering space $\tilde{M}$ of a closed 4-manifold M is homotopy equivalent to S^3 if and only if $\pi_1(M)$ has two ends and $\chi(M) = 0$. In particular, $\pi = \pi_1(M)$ has a maximal finite normal subgroup F, with cohomological period dividing 4, and the quotient π/F is isomorphic to Z or D. (Thus there is a subgroup of index at most 2 which is a semidirect product $F \times_\theta Z$). Moreover, M is determined up to homotopy equivalence by π and the orbit of the first nontrivial k-invariant under the action of $Aut(\pi)$.

Theorem 1. *Let M be a closed 4-manifold with fundamental group π. Then*

Typeset by $\mathcal{A}_{\mathcal{M}}\mathcal{S}$-TeX

M is virtually an $S^3 \times E^1$-manifold if and only if $\chi(M) = 0$ and π has two ends. If these conditions hold then
(i) M is finitely covered by $S^3 \times S^1$; and
(ii) The covering space M_F corresponding to a finite normal subgroup of $\pi_1(M)$ has the homotopy type of an orientable finite PD_3-complex.
If M is itself an $S^3 \times E^1$-manifold then F is the fundamental group of an S^3-manifold.

Proof. An orientable manifold is an $S^3 \times E^1$-manifold if and only if it is homeomorphic to the mapping torus of an orientation preserving self homeomorphism of an S^3-manifold, by Theorem A(1) of [Ue91]. Thus the conditions are clearly necessary.

Suppose that they hold. Let $\hat{M}$ be a finite covering space of M which is orientable and has fundamental group Z. The universal covering space $\tilde{M}$ is homotopy equivalent to S^3, by Theorem II.10. Since $\hat{M}$ is orientable the generator of the group of covering translations $Aut(\tilde{M}/\hat{M}) \cong Z$ is homotopic to the identity. Therefore $\hat{M} \simeq \tilde{M} \times S^1 \simeq S^3 \times S^1$, and so is homeomorphic to $S^3 \times S^1$, by surgery over Z. Hence M is virtually an $S^3 \times E^1$-manifold and $\tilde{M}$ is homeomorphic to $S^3 \times R$.

The covering space $M_F = \tilde{M}/F$ is an orientable PD_3-complex since $\tilde{M} \simeq S^3$. (See Theorem II.10 or Theorem 4.3 of [Wl67]). The image of the finiteness obstruction for M_F under the "geometrically significant injection" of $K_0(Z[F])$ into $Wh(F \times Z)$ of [Rn86] is the obstruction to $M_F \times S^1$ being a simple PD-complex. Since the group of self homotopy equivalences of M_F is finite [Pl82] M has a finite covering space which is homotopy equivalent to $M_F \times S^1$ and so this PD_4-complex is simple. Therefore M_F is finite.

The final assertion follows from Theorem A(1) of [Ue91] as π has a subgroup of finite index which is isomorphic to $F \times Z$ and which acts orientably on $\tilde{M}$. //

Corollary. *The universal covering space of M is homeomorphic to $S^3 \times R \simeq R^4\backslash\{0\}$, and F acts trivially on $\pi_3(M) \cong Z$.* //

2. The maximal finite normal subgroup

The action of π by conjugation on F induces a homomorphism from π/F to the group of outer automorphisms $Out(F)$.

Lemma 2. *The quotient π/F acts on $\pi_3(M)$ and $H^4(F;Z)$ through multiplication by ± 1. It acts trivially if the order of F is divisible by 4 or by any prime congruent to 3 modulo (4).*

Proof. The group π/F must act through ± 1 on the infinite cyclic groups $\pi_3(M)$ and $H_3(M_F;Z)$. By the universal coefficient theorem $H^4(F;Z)$ is isomorphic to $H_3(F;Z)$, which is the cokernel of the Hurewicz homomorphism from $\pi_3(M)$ to $H_3(M_F;Z) \cong Z/|F|Z$, and so the first assertion holds. To prove the second assertion we may pass to the Sylow subgroups of F, by Shapiro's Lemma. If p is an odd prime then the p-Sylow subgroups of F are cyclic, while the 2-Sylow subgroups of F are cyclic or generalized quaternionic. In all cases an automorphism induces multiplication by a square on the third homology. But -1 is not a square modulo 4 nor modulo any prime $p = 4n+3$. //

Let $I(F)$ denote the subgroup of $Out(F)$ which acts by ± 1 on $H_3(F;Z)$, and let $I_+(F)$ be the subgroup which acts trivially. The list of finite groups with cohomological period dividing 4 is well known (see [DM85]). In determining the outer automorphism groups $Out(F)$ and their subgroups $I(F)$ for these groups we have used the following simple facts: (i) an outer automorphism class induces multiplication by r on $H^4(F;Z)$ if and only if it does so for each Sylow subgroup of F, by Shapiro's Lemma; (ii) automorphisms of $Q(2^n)$ induce multiplication by squares on $H^4(Q(2^n);Z)$ [Sw60].

If m is an integer let $l(m)$ be the number of odd prime divisors of m.

(i) Z/dZ: $Aut(Z/dZ) = Out(Z/dZ) = (Z/dZ)^\times$. $I(Z/dZ) = \{s \in (Z/dZ)^\times | s^2 = \pm 1\}$. $I_+(Z/dZ) = (Z/2Z)^{l(d)}$ if $d \not\equiv 0(4)$, $(Z/2Z)^{l(d)+1}$ if $d \equiv 4(8)$, and $(Z/2Z)^{l(d)+2}$ if $d \equiv 0(8)$.

(ii) $Q = Q(8) =< x, y | x^2 = y^2 = (xy)^2 >$: $Aut(Q)$ is the semidirect product of $Aut(Q/Q') \cong SL(2,2)$ with the normal subgroup $Inn(Q) = Q/Q' \cong (Z/2Z)^2$. $I(Q) = Out(Q) \cong SL(2,2) \cong S_3$, generated by the images of the automorphisms σ which sends x and y to y and xy, respectively, and τ which interchanges x and y.

(iii) $Q(8k) =< x, y | x^{4k} = 1, x^{2k} = y^2, yxy^{-1} = x^{-1} >$, where $k > 1$: $Aut(Q(8k))$ is generated by the automorphisms (i, s) which send x and y to x^s and $x^i y$, respectively, where $(s, 2k) = 1$, and is isomorphic to the semidirect product of $(Z/4kZ)^\times$ with the normal subgroup $Z/4kZ$. $Out(Q(8k)) = (Z/2Z) \oplus ((Z/4kZ)^\times/(\pm 1))$, generated by the images of the $(0, s)$ and $(1,1)$. $I(Q(8k)) = (Z/2Z)^{l(k)+1}$ if k is odd and $(Z/2Z)^{l(k)+2}$ if k is even.

(iv) $T_k^* =< Q, z | z^{3^k} = 1, zxz^{-1} = y, zyz^{-1} = xy >$, where $k \geq 1$: a presentation for $Aut(T_k^*)$ is given in [H]. $Out(T_k^*) = (Z/3^k Z)^\times$ is cyclic of order 2.3^{k-1} and is generated by the image of the automorphism ρ which sends x, y and z to y^{-1}, x^{-1} and z^2 respectively. $I(T_k^*) = Z/2Z$.

(v) $O_k^* =< T_k^*, w | w^2 = x^2, wxw^{-1} = yx, wyw^{-1} = y^{-1}, wzw^{-1} = z^{-1} >$, where $k \geq 1$: the automorphism ρ of T_k^* extends to O_k^* via $\rho(w) = w^{-1}z^2$ and so the natural homomorphism from $Out(O_k^*)$ to $Out(T_k^*)$ is onto. Its kernel is generated by the image of the automorphism which sends w to wx^2 and is the identity on the commutator subgroup T_k^*. This automorphism induces multiplication by 25 on the third homology of the 2-Sylow subgroup $Q(16)$ of O_v^*. The generator $\rho^{3^{k-1}}$ of $I(T_k^*)$ is induced by conjugation by wz in O_k^* and so is the restriction of an inner automorphism of O_k^*. $I(O_k^*) = 1$.

(vi) $I^* =< x, y | x^2 = y^3 = (xy)^5 >$: $Aut(I^*) \cong S_5$ and $Out(I^*) = Z/2Z$; the nontrivial outer automorphism induces multiplication by 49 on $H_3(I^*; Z)$ [Pl82]. $I(I^*) = 1$.

(vii) $A(m, e) =< x, y | x^m = y^{2^e} = 1, yxy^{-1} = x^{-1} >$, where $m > 1$ is odd and $e \geq 1$: $Aut(A(m, e))$ is generated by the automorphisms (s, t, u) which send x and y to x^s and $x^u y^t$, respectively, where $(s, m) = (t, 2) = 1$. $Out(A(m, e))$ is generated by the images of $(s, 1, 0)$ and $(1, t, 0)$ and is isomorphic to $(Z/2^e)^\times \oplus ((Z/mZ)^\times/(\pm 1))$. $I(A(m, 1)) = \{s \in (Z/mZ)^\times | s^2 = \pm 1\}/(\pm 1)$, $I(A(m, 2)) = (Z/2Z)^{l(m)}$, $I(A(m, e)) = (Z/2Z)^{l(m)+1}$ if $e > 2$.

(viii) $Q(2^n a, b, c) =< Q(2^n), u | u^{abc} = 1, xu^{ab} = u^{ab}x, xu^c x^{-1} = u^{-c}, yu^{ac} = u^{ac}y, yu^b y^{-1} = u^{-b} >$, where a, b and c are odd and relatively prime, at most one of a, b and c is 1 and $n > 2$: there is a natural homomorphism from $Aut(Q(2^n a, b, c))$ to $Aut(Q(2^n)) \times Aut(Z/abcZ)$ with kernel consisting of inner automorphisms. Hence $Out(Q(2^n a,b,c))$ is a subquotient of $Out(Q(2^n)) \times (Z/abcZ)^\times$. $I(Q(2^n a,b,c))$ has exponent 2, even if $n > 3$.

In addition there are the direct products of any of these groups with a cyclic group of relatively prime order. As $Aut(G \times H) = Aut(G) \times Aut(H)$ and $Out(G \times H) = Out(G) \times Out(H)$ if G and H are finite groups of relatively prime order, we have $I_+(G \times Z/dZ) = I_+(G) \times I_+(Z/dZ)$. In particular, if G is not cyclic or dihedral then $I(G \times Z/dZ) = I_+(G \times Z/dZ) = I(G) \times I_+(Z/dZ)$.

In all cases except when F is cyclic or $Q \times Z/dZ$ the group $I(F)$ has exponent 2 and hence π has a subgroup of index at most 4 which is isomorphic to $F \times Z$.

3. Extensions of D

We shall now assume that $\pi/F \cong D$. Let $u, v \in D$ be a pair of involutions which generate D and let $s = uv$. Then $s^{-n}us^n = us^{2n}$, and any involution in D is conjugate to u or to $v = us$. Hence any pair of involutions $\{u', v'\}$ which generates D is conjugate to the pair $\{u, v\}$, up to change of order.

Theorem 3. *Let M be a closed 4-manifold with $\chi(M) = 0$, and such that there is an epimorphism $p : \pi \to D$ with finite kernel F, and let $\hat{u}$ and $\hat{v}$ be a pair of elements of π whose images $u = p(\hat{u})$ and $v = p(\hat{v})$ are involutions which together generate D. Then*

(i) M is nonorientable and $\hat{u}, \hat{v}$ each represent orientation reversing loops;

(ii) the subgroups G and H generated by F and $\hat{u}$ and by F and $\hat{v}$, respectively, each have cohomological period dividing 4, and the unordered pair $\{G, H\}$ of groups is determined up to isomorphisms by π alone;

(iii) conversely, π is determined up to isomorphism by the unordered pair $\{G, H\}$ of groups with index 2 subgroups isomorphic to F as the free product with amalgamation $\pi = G *_F H$;

(iv) π acts trivially on $\pi_3(M)$.

Proof. The first assertion follows from Lemma II.14, since π/π' is finite. Let $\hat{s} = \hat{u}\hat{v}$. Suppose that $\hat{u}$ is orientation preserving. Then the subgroup σ generated by $\hat{u}$ and $\hat{s}^2$ is orientation preserving so the corresponding covering space M_σ is orientable. But σ has finite index in π and σ/σ' is finite and so this contradicts Lemma II.14. Similarly, $\hat{v}$ must be orientation reversing.

By assumption, $\hat{u}^2$ and $\hat{v}^2$ are in F, and $[G : F] = [H : F] = 2$. If F is not isomorphic to $Q \times Z/dZ$ then $I(F)$ is abelian and so the (normal) subgroup generated by F and $\hat{s}^2$ is isomorphic to $F \times Z$. In any case the subgroup generated by F and $\hat{s}^k$ is normal, and is isomorphic to $F \times Z$ if k is a nonzero multiple of 12. The uniqueness up to isomorphisms of the pair $\{G, H\}$ follows from the uniqueness up to conjugation and order of the pair of generating involutions for D. Since G and H act freely on $\tilde{M}$ they also have cohomological period dividing 4. On examining the list above we see that F must be cyclic or the product of $Q(8k)$, T_v^* or $A(m,e)$ with a cyclic group of relatively prime order, as it is the kernel of a map from G to $Z/2Z$. It is easily verified that in all such cases every automorphism of F is the restriction of automorphisms of G and H. It follows that π is determined up to isomorphism as the amalgamated free product $G *_F H$ by the unordered pair $\{G, H\}$ of groups with index 2 subgroups isomorphic to F (i.e., it is unnecessary to specify the identifications of F with these subgroups).

The final assertion follows because each of the spaces $M_G = \tilde{M}/G$ and $M_H = \tilde{M}/H$ are PD_3-complexes with finite fundamental group and therefore are orientable, and π is generated by G and H. //

Must the covering spaces M_G and M_H be homotopy equivalent to finite complexes?

4. Realization of the groups

Each finite group with cohomological period dividing 4 is the fundamental group of some finitely dominated PD_3-complex [Sw60]. The homotopy type of such a "Swan complex" for F is determined by a k-invariant which is a generator of $H^3(F; Z) \cong Z/|F|Z$. Thus Swan complexes for F are parametrized up to homotopy by the quotient of $(Z/|F|Z)^\times$ under the action of $Out(F)$) and $Aut(Z)$. The set of finiteness obstructions for all such complexes forms a coset of the Swan subgroup T of $\tilde{K}_0(Z[F])$. Thus there is a finite complex of this type if an obstruction $\sigma_4(F)$ in $\tilde{K}_0(Z[F])/T$ is 0. (This obstruction is nonzero if F has a subgroup isomorphic to O_k^* for some $k > 1$ [DM85]). We shall henceforth let X_F denote a finite Swan complex for F.

All the groups listed above except the groups O_k^* with $k > 1$, $A(m,1)$ and $Q(2^n a, b, c)$ and their products with cyclic groups have fixed point free representations in $SO(4)$ and so act linearly on S^3. As the corresponding quotient spaces are the 3-dimensional Clifford-Klein space forms we shall call such groups CK groups. If F is cyclic then every Swan complex for F is homotopy equivalent to a lens space. If $F = Q(2^k)$ or T_k^* for some $k > 1$ then S^3/F is the unique finite Swan complex for F [Th80]. For the other noncyclic CK groups there is an unique spherical space form, but in general there are other finite Swan complexes.

The map from X_F to BG classifying the Spivak normal fibration of X_F lifts to a map from X_F to $BTOP$, and so there are normal maps $(f,b) : N^3 \to X_F$ [MTW76]. Given such a normal map there is a "proper surgery" obstruction $\lambda^p(f,b)$ in $L_3^p(F)$ which is 0 if and only if $(f,b) \times id_{S^1}$ is normally cobordant to a simple homotopy equivalence. In [HM86] it is shown that if $F \times Z$ acts freely and properly on $R^{4d}\backslash\{0\}$, where F is a finite group of cohomological period dividing $4d$, then a surgery semicharacteristic must be 0. In particular, F has no subgroup isomorphic to $A(m,1)$ (with $m > 1$) or $Q(2^n a, b, c)$ (with $n > 3$ and b or $c > 1$). Thus we may focus on the cases when $F \cong Q(8a,b,c) \times Z/dZ$, where a, b and c are odd and at most one of them is 1. The main result of [HM86] is that in such a case $F \times Z$ acts freely and properly "with almost linear k-invariant" if and only if some arithmetical conditions depending on subgroups of F of the form $Q(8a,b,1)$ hold. (The high dimensional surgery arguments of [MTW76] and [HM86] may be extended to the 4-dimensional case by [FQ]).

If F acts freely on an homology 3-sphere Σ there is a more direct argument for the existence of a free proper action of $F \times Z$ on $S^3 \times R$. (The following argument was outlined in [KS88]). Let $\Pi = \pi_1(\Sigma/F)$ and lift a cellular decomposition of Σ/F to equivariant cellular decompositions of Σ and its universal covering space $\tilde{\Sigma}$. Then $C_*(\Sigma) = F \otimes_\Pi C_*(\tilde{\Sigma})$ is a finitely generated free periodic resolution of Z over $Z[F]$. Let X be the corresponding finite Swan complex. Since the spaces involved are 3-dimensional there is a map $h : \Sigma/F \to X$ realizing the chain map (over the epimorphism : $\Pi \to F$) from $C_*(\tilde{\Sigma})$ to $C_*(\tilde{X})$. Since the map h is a $Z[F]$-homology equivalence the product $h \times id_{S^1}$ is a simple $Z[F \times Z]$-homology equivalence and so has surgery obstruction 0 in $L_4^s(F \times Z)$. Therefore we may do surgery to obtain a simple homotopy equivalence.

For example, the results of [HM86] imply that there is a closed orientable 4-manifold M which is simple homotopy equivalent to a product $X_F \times S^1$ where $F = Q(24, 13, 1)$. If $Q(24, 13, 1)$ is not a 3-manifold group then M cannot fibre over S^1, and so we would have a counter example to a 4-dimensional analogue of the Farrell fibration theorem of a different kind from that of [We87]. In any case this group is not isomorphic to a subgroup of $SO(4)$ which acts freely on S^3 and so M is not geometric. (Can one give an explicit example of a free action of $Q(24, 13, 1)$ on an homology 3-sphere?)

If $\pi/F \cong Z$ and $\tau : M_F \to M_F$ is a generator of the group of covering transformations then M is homotopy equivalent to the mapping torus $M(\tau)$. Conversely, by Corollary 1.3 of [Pl82] the map sending a self homotopy equivalence h of X_F to the induced outer automorphism class $[\theta]$ determines an isomorphism from the group of homotopy classes of self homotopy equivalences $E(X_F)$ to $I(F)$, provided that $|F| > 2$. By Lemma 2 above, if 4 divides $|F|$ then h is orientation preserving. The mapping torus $M(h)$ has fundamental group $F \times_\theta Z$ and is a finite PD_4-complex, by Proposition 4.1 of [Ra86]. Moreover, if $\pi \cong F \times Z$ and $|F| > 2$ then h is homotopic to the identity and so $M(h)$ is homotopy equivalent to $X_F \times S^1$. (There is a minor oversight in [Pl82]. If $F = 1$ or $Z/2Z$ then X_F admits an orientation reversing involution which induces the identity on F).

If $F = Z/dZ$ then the automorphism of F corresponding to an integer s such that $s^2 \equiv \pm 1(d)$ is realizable by an isometry of the lens space $L(d, s)$, and the semidirect product $(Z/dZ) \times_{(s)} Z$ is the fundamental group of the mapping torus. If $d > 2$ a closed 4-manifold with this group and with Euler characteristic 0 is orientable if and only if $s^2 \equiv 1(d)$. If F is Q, T_k^*, O^*, I^*, $A(p^i, e)$, $Q \times Z/q^j Z$ or $A(p^i, 2) \times Z/q^j Z$ where p and q are odd primes and $e > 1$ then every class in $I(F)$ is realizable by an isometry of S^3/F [Rb79]. The mapping torus of such an isometry is an orientable $S^3 \times E^1$-manifold. For the other CK groups the subgroup of $I(F)$ realizable by homeomorphisms of S^3/F is usually quite small (cf. [BR84, HR83, Rb79, Ue91]). Is the discrepancy to be explained by the fact that a closed 4-manifold is a *simple* PD_4-complex?

If F, G and H are each noncyclic CK groups then the corresponding spherical space forms are uniquely determined, and we may construct a nonorientable $S^3 \times E^1$-manifold with fundamental group $\pi = G *_F H$ as follows. Let u and $v : S^3/F \to S^3/F$ be the covering involutions with quotient spaces S^3/G and S^3/H, respectively, and let $\phi = uv$. (Note that u and v are isometries of S^3/F). Then $U([x,t]) = [u(x), 1-t]$ defines a fixed point free involution on the mapping torus $M(\phi)$ and the fundamental group of the quotient space is isomorphic to $G *_F H$. If F is cyclic and $G \cong H$ or if G is cyclic then a similar construction works, but in general if F is cyclic the covering spaces of S^3/G and S^3/H with group F may be distinct lens spaces.

As recalled above the homotopy type of M is determined by π and the class of the first nontrivial k-invariant, which is an element of $H^4(\pi; \pi_3(M))$, modulo the action of $Aut(\pi)$. If $\pi/F \cong Z$ then the k-invariant is a generator of $H^4(\pi; \pi_3(M)) \cong H^4(F; Z) \cong Z/|F|Z$, while if $\pi = G *_F H$ then $H^4(\pi; Z) \cong \{(\zeta, \xi) \in Z/|G|Z \oplus Z/|H|Z : \zeta \equiv \xi \bmod (|F|)\} \cong Z/2|F|Z \oplus Z/2Z$, and the k-invariant restricts to a generator of each of the groups $H^4(G; Z)$ and $H^4(H; Z)$. In particular, if $\pi \cong D$ (or Z) the k-invariant is unique, and so any closed 4-manifold M with $\pi_1(M) \cong D$ and $\chi(M) = 0$ is homotopy equivalent to $RP^4 \sharp RP^4$.

In [HM86] it is shown that the group $Q(8k) \times Z/dZ \times Z$ can only act freely and properly on $R^4 \backslash \{0\}$ with the k-invariant corresponding to the free linear action of $Q(8k) \times Z/dZ$ on S^3. However in general it is not known which k-invariants are realizable. (Every group of the form $Q(8a, b, c) \times Z/dZ \times Z$ admits an "almost linear" k-invariant, but there may be other actions. Further results on this question may be found in [HM86']).

In Chapter 8 of [H] it is shown that if M is orientable, $\pi/\pi' \cong Z$ and $F = \pi'$ is a nontrivial finite group then the structure set $S_{TOP}(M)$ is infinite. As the group of self homotopy equivalences of such a manifold is finite, by Theorem VII.2 of [H], it follows that there are infinitely many distinct topological 4-manifolds simple homotopy equivalent to M. For instance, as $Wh(Z \oplus (Z/2Z)) = 0$ [Kw86] and $L_5(Z \oplus (Z/2Z), +) \cong Z^2$, by Theorem 13A.8 of [W], the set $S_{TOP}(RP^3 \times S^1)$ is infinite. Although all of the manifolds in this homotopy type are doubly covered by $S^3 \times S^1$ only $RP^3 \times S^1$ is itself geometric. It is likely that similar estimates hold for the other manifolds covered by $S^3 \times R$ (if $\pi \neq Z$) and we shall not pursue the question of homeomorphism types further.

CHAPTER IX

GEOMETRIES WITH COMPACT MODELS

There are three geometries with compact models, namely S^4, CP^2 and $S^2 \times S^2$. The first two of these are easily dealt with, as there is only one other geometric manifold, namely RP^4, and for each of the two projective spaces there is one other (nonsmoothable) manifold of the same homotopy type. With the geometry $S^2 \times S^2$ we shall consider also the bundle space $S^2 \tilde{\times} S^2$. There are eight $S^2 \times S^2$-manifolds, seven of which are total spaces of bundles with base and fibre each S^2 or RP^2, and there are two other such bundle spaces covered by $S^2 \tilde{\times} S^2$.

The universal covering space $\tilde{M}$ of a closed 4-manifold M is homeomorphic to $S^2 \times S^2$ if and only if $\pi = \pi_1(M)$ is finite, $\chi(M)|\pi| = 4$ and $w_2(\tilde{M}) = 0$. (The condition $w_2(\tilde{M}) = 0$ may be restated entirely in terms of M, but at somewhat greater length). If these conditions hold and π is cyclic then M is homotopy equivalent to an $S^2 \times S^2$-manifold, except when $\pi = Z/2Z$ and M is nonorientable, in which case there is one other homotopy type. However we have not been able to characterize completely the possible k-invariants when $\pi \cong (Z/2Z)^2$. Moreover, there is a further ambiguity of order at most 4 in determining the homotopy type. If $\chi(M)|\pi| = 4$ and $w_2(\tilde{M}) \neq 0$ then either $\pi = 1$, in which case $M \simeq S^2 \tilde{\times} S^2$ or $CP^2 \sharp CP^2$, or M is nonorientable and $\pi = Z/2Z$; in the latter case $M \simeq RP^4 \sharp CP^2$, the nontrivial RP^2-bundle over S^2, and $\tilde{M} \simeq S^2 \tilde{\times} S^2$.

The number of homeomorphism classes within each homotopy type is at most two if $\pi = Z/2Z$ and M is orientable, two if $\pi = Z/2Z$, M is nonorientable and $w_2(\tilde{M}) = 0$, four if $\pi = Z/2Z$ and $w_2(\tilde{M}) \neq 0$, at most four if $\pi \cong Z/4Z$, and at most eight if $\pi \cong (Z/2Z)^2$. We do not know whether there are enough exotic self homotopy equivalences to account for all the normal invariants with trivial surgery obstruction. However a PL 4-manifold with

Typeset by $\mathcal{A}\mathcal{M}\mathcal{S}$-TeX

the same homotopy type as a geometric manifold or $S^2 \tilde{\times} S^2$ is homeomorphic to it, in (at least) nine of the 13 cases. (In seven of these cases the homotopy type is determined by the Euler characteristic, fundamental group and Stiefel-Whitney classes).

For the full details of some of the arguments in the cases $\pi \cong Z/2Z$ we refer to the forthcoming papers [KKR92] and [HKT92].

1. The geometries S^4 and CP^2.

The unique element of $Isom(S^4) = O(5)$ of order 2 which acts freely on S^4 is $-I$. Therefore S^4 and RP^4 are the only S^4-manifolds. The manifold S^4 is determined up to homeomorphism by the conditions $\chi(S^4) = 2$ and $\pi_1(S^4) = 1$ [FQ].

Lemma 1. [Ol53] *A closed 4-manifold M is homotopy equivalent to RP^4 if and only if $\chi(M) = 1$ and $\pi_1(M) = Z/2Z$.*

Proof. The conditions are clearly necessary. Suppose that they hold. Then $\tilde{M} \simeq S^4$ and $w_1(M) = w_1(RP^4) = w$, say, since any orientation preserving self homeomorphism of $\tilde{M}$ has Lefshetz number 2. Since $RP^\infty = K(Z/2Z, 1)$ may be obtained from RP^4 by adjoining cells of dimension at least 5 we may assume $c_M = c_{RP^4} f$, where $f : M \to RP^4$. Since c_{RP^4} and c_M are each 4-connected f induces isomorphisms on homology with coefficients $Z/2Z$. Considering the exact sequence of homology corresponding to the short exact sequence of coefficients $0 \to Z^w \to Z^w \to Z/2Z \to 0$, we see that f has odd degree. By modifying f on a 4-cell $D^4 \subset M$ we may arrange that f has degree 1, and the lemma then follows from Theorem II.3. //

This lemma may also be proven by comparison of the k-invariants of M and RP^4, as in Theorem 4.3 of [Wl67].

By Theorems 13.A.1 and 13.B.5 of [W] the surgery obstruction homomorphism is determined by an Arf invariant and maps $[RP^4; G/TOP]$ onto $Z/2Z$, and hence the structure set $S_{TOP}(RP^4)$ has two elements. (See the discussion of nonorientable manifolds with fundamental group $Z/2Z$ in Section 6 below for more details). As every self homotopy equivalence of RP^4 is homotopic to the identity [Ol53] there is one fake RP^4. The fake RP^4 is not smoothable [Ru84].

There is a similar characterization of the homotopy type of the complex projective plane.

Lemma 2. *A closed 4-manifold M is homotopy equivalent to CP^2 if and only if $\chi(M) = 3$ and $\pi_1(M) = 1$.*

Proof. The conditions are clearly necessary. Suppose that they hold. Then $H^2(M;Z)$ is infinite cyclic and so there is a map $f_M : M \to CP^\infty = K(Z,2)$ which induces an isomorphism on H^2. Since CP^∞ may be obtained from CP^2 by adjoining cells of dimension at least 6 we may assume $f_M = f_{CP^2}g$, where $g : M \to CP^2$ and $f_{CP^2} : CP^2 \to CP^\infty$ is the natural inclusion. As $H^4(M;Z)$ is generated by $H^2(M;Z)$, by Poincaré duality, g induces an isomorphism on cohomology and hence is a homotopy equivalence. //

In this case the surgery obstruction homomorphism is determined by the difference of signatures and maps $[CP^2; G/TOP]$ onto Z. The structure set $S_{TOP}(CP^2)$ again has two elements, and as every self homotopy equivalence of CP^2 is homotopic to the identity there is one fake CP^2. The fake CP^2 is also known as the Chern manifold Ch or $*CP^2$, and is not smoothable [FQ]. Neither of these manifolds admits a nontrivial fixed point free action, as any self map of CP^2 or Ch has nonzero Lefshetz number, and so CP^2 is the only CP^2-manifold.

2. The geometry $S^2 \times S^2$.

The manifold $S^2 \times S^2$ is determined up to homotopy equivalence by the conditions $\chi(S^2 \times S^2) = 4$, $\pi_1(S^2 \times S^2) = 1$ and $w_2(S^2 \times S^2) = 0$, by Theorem IV.7. These conditions in fact determine $S^2 \times S^2$ up to homeomorphism [FQ]. Hence if M is an $S^2 \times S^2$-manifold its fundamental group π is finite, $\chi(M)|\pi| = 4$ and $w_2(\tilde{M}) = 0$.

The isometry group of $S^2 \times S^2$ is generated by the elements τ, where $\tau(x,y) = (y,x)$, and (A,B), where $A, B \in O(3)$, and is the semidirect product $(O(3) \times O(3))\tilde{\times}(Z/2Z)$. The involution τ acts on $O(3) \times O(3)$ by $\tau(A,B)\tau = (B,A)$ for $A, B \in O(3)$. In particular, $(\tau(A,B))^2 = id$ if and only if $AB = I$, and so such an involution fixes (x, Ax), for any $x \in S^2$. Thus there are no free $Z/2Z$-actions in which the factors are switched. The element (A, B) generates a free $Z/2Z$-action if and only if $A^2 = B^2 = I$ and at least one of A, B acts freely, i.e. if A or $B = -I$. After conjugation with τ if necessary we may assume that $B = -I$, and so there are four conjugacy classes in $Isom(S^2 \times S^2)$ of free $Z/2Z$-actions. (The conjugacy classes may be distinguished by the multiplicity (0, 1, 2 or 3) of 1 as an eigenvalue of A). In each case the projection onto the second factor gives rise to a fibre bundle projection from the orbit space to RP^2, with fibre S^2.

If the involutions (A, B) and (C, D) generate a free $(Z/2Z)^2$-action then (AC, BD) is also a free involution. By the above paragraph, one element of each of these ordered pairs must be $-I$. It follows easily that (after conjugation with τ if necessary) the $(Z/2Z)^2$-actions are generated by pairs $(A, -I)$ and $(-I, I)$, where $A^2 = I$. Since A and $-A$ give rise to the same subgroup, there are two free $(Z/2Z)^2$-actions. The orbit spaces are the total spaces of RP^2-bundles over RP^2.

If $(\tau(A, B))^4 = id$ then (BA, AB) is a fixed point free involution and so $BA = AB = -I$. Since $(A, I)\tau(A, -A^{-1})(A, I)^{-1} = \tau(I, -I)$ every free $Z/4Z$-action is conjugate to the one generated by $\tau(I, -I)$. The orbit space does not fibre over a surface. (See below).

In the next section we shall see that these eight geometric manifolds may be distinguished by their fundamental group and Stiefel-Whitney classes. Note that if F is a finite group then $q(F) \geq 2/|F| > 0$, while $q^{SG}(F) \geq 2$. Thus S^4, RP^4 and the geometric manifolds with $|\pi| = 4$ have minimal Euler characteristic for their fundamental groups (i.e., $\chi(M) = q(\pi)$), while $S^2 \times S^2/(-I, -I)$ has minimal Euler characteristic among orientable PD_4-complexes realizing $Z/2Z$.

3. Bundle spaces.

There are two S^2-bundles over S^2, since $\pi_1(SO(3)) = Z/2Z$. The total space $S^2 \tilde{\times} S^2$ of the nontrivial S^2-bundle over S^2 is determined up to homotopy equivalence by the conditions $\chi(S^2 \tilde{\times} S^2) = 4$, $\pi_1(S^2 \tilde{\times} S^2) = 1$, $w_2(S^2 \tilde{\times} S^2) \neq 0$ and $\sigma(S^2 \tilde{\times} S^2) = 0$, by Theorem IV.7. However there is one fake $S^2\tilde{\times}S^2$. The bundle space is homeomorphic to the connected sum $CP^2\sharp -CP^2$, while the fake version is homeomorphic to $CP^2\sharp -Ch$ and is not smoothable [FQ]. There are two other closed 4-manifolds with $\pi = 1$, $\chi = 4$ and $w_2 \neq 0$ but with nonzero signature, namely $CP^2\sharp CP^2$ and $CP^2\sharp Ch$. However we shall see that they do not properly cover any other 4-manifold.

Since the Kirby-Siebenmann obstruction of a closed 4-manifold is natural with respect to covering maps and dies on passage to 2-fold coverings, the nonsmoothable manifolds $CP^2\sharp -Ch$ and $CP^2\sharp Ch$ admit no nontrivial free involution. The following lemma implies that the spaces $S^2\tilde{\times}S^2$ and $CP^2\sharp CP^2$ admit no orientation preserving free involution, and hence no free action of $Z/4Z$ or $(Z/2Z)^2$.

Lemma 3. *Let M be a closed 4-manifold with fundamental group $\pi = Z/2Z$ and universal covering space $\tilde{M}$. Then*
(i) $w_2(\tilde{M}) = 0$ if and only if $w_2(M) = u^2$ for some $u \in H^1(M; Z/2Z)$; and
(ii) if M is orientable and $\chi(M) = 2$ then $w_2(\tilde{M}) = 0$ and so $\tilde{M} \cong S^2 \times S^2$.
Proof. The Cartan-Leray cohomology spectral sequence (with coefficients $Z/2Z$) for the projection $p : \tilde{M} \to M$ gives an exact sequence

$$0 \to H^2(\pi; Z/2Z) \to H^2(M; Z/2Z) \to H^2(\tilde{M}; Z/2Z),$$

in which the right hand map is induced by p and has image in the subgroup fixed under the action of π. Hence $w_2(\tilde{M}) = p^* w_2(M)$ is 0 if and only if $w_2(M)$ is in the image of $H^2(\pi; Z/2Z)$. Since $\pi = Z/2Z$ this is so if and only if $w_2(M) = u^2$ for some $u \in H^1(M; Z/2Z)$.

Suppose now that M is orientable and $\chi(M) = 2$. Then $H^2(\pi; Z) = H^2(M; Z) = Z/2Z$. Let x generate $H^2(M; Z)$ and let $\bar{x}$ be its image under reduction modulo (2) in $H^2(M; Z/2Z)$. Then $\bar{x} \cup \bar{x} = 0$ in $H^4(M; Z/2Z)$ since $x \cup x = 0$ in $H^4(M; Z)$. Moreover as M is orientable $w_2(M) \cup \bar{x} = \bar{x} \cup \bar{x}$, by the Wu formula, and so is 0. Since the cup product pairing on $H^2(M; Z/2Z) \cong (Z/2Z)^2$ is nondegenerate it follows that $w_2(M) = \bar{x}$ or 0. Hence $w_2(\tilde{M})$ is the reduction of $p^* x$ or is 0. The integral analogue of the above exact sequence implies that the natural map from $H^2(\pi; Z)$ to $H^2(M; Z)$ is an isomorphism and so $p^*(H^2(M; Z)) = 0$. Hence $w_2(\tilde{M}) = 0$ and so $\tilde{M} \cong S^2 \times S^2$. //

Since $RP^2 = Mb \cup D$ is the union of a Möbius band Mb and a disc D, an F-bundle over RP^2 is determined by a bundle over Mb which restricts to a trivial bundle over ∂Mb, i.e. by a conjugacy class of elements of order dividing 2 in $\pi_0(Homeo(F))$, together with the class of a gluing map over $\partial Mb = \partial D$ modulo those which extend across D or Mb, i.e. an element of the quotient $\pi_1(Homeo(F))/(squares)$. Hence there are four S^2-bundles over RP^2.

The orbit space $M = (S^2 \times S^2)/(A, -I)$ is orientable if and only if $det(A) = -1$. If A has a fixed point $P \in S^2$ then the image of $\{P\} \times S^2$ in M is an embedded projective plane which represents a nonzero class in $H_2(M; Z/2Z)$. If $A = I$ or is a reflection across a plane the fixed point set has dimension > 0 and so this projective plane has self intersection 0. As the fibre S^2 intersects this projective plane in one point and has self intersection 0 it follows that the second Wu class $v_2(M) = w_2(M) + w_1(M)^2$ is 0 and so $w_2(M) = w_1(M)^2$ in these two cases. If A is a rotation about an axis then the projective plane has self intersection 1. Finally, if $A = -I$ then the image of the diagonal $\{(x, x) | x \in S^2\}$ is a projective plane in M with self intersection 1 [Ha82]. Thus in these two cases $v_2(M) \neq 0$. Therefore, by part *(i)* of the lemma, $w_2(M)$ is the square of the nonzero element of $H^1(M; Z/2Z)$ if $A = -I$ and is 0 if A is a rotation. Thus these bundle spaces may be distinguished by their Stiefel-Whitney classes, and every S^2-bundle over RP^2 is geometric.

The group $E(RP^2)$ of self homotopy equivalences of RP^2 is connected and the natural map from $SO(3)$ to $E(RP^2)$ induces an isomorphism on π_1, by Lemma IV.2. Hence there are two RP^2-bundles over S^2, up to fibre homotopy equivalence. The total space of the nontrivial RP^2-bundle over S^2 is the quotient of $S^2 \tilde{\times} S^2$ by the bundle involution which is the antipodal map on each fibre. If we observe that $S^2 \tilde{\times} S^2 \cong CP^2 \sharp - CP^2$ is the union of two copies of the D^2-bundle which is the mapping cone of the Hopf fibration and that this involution interchanges the hemispheres we see that this space is homeomorphic to $RP^4 \sharp CP^2$.

There are two RP^2-bundles over RP^2. (The total spaces of each of the latter bundles have fundamental group $(Z/2Z)^2$, since $w_1 : \pi \rightarrow \pi_1(RP^2) = Z/2Z$ restricts nontrivially to the fibre, and so is a splitting homomorphism for the homomorphism induced by the inclusion of the fibre). They may be distinguished by their orientation double covers, and each is geometric.

4. The homotopy type - the action of π on $\pi_2(M)$.

We shall assume henceforth that M is a closed connected 4-manifold with finite fundamental group π and that $\chi(M)|\pi| = 4$. Moreover we may choose a basis for $\pi_2(M) \cong Z^2$ so that the corresponding basis of the intersection form on $\pi_2(M)$ is J or $\tilde{J}$, as in Section 4. The *quadratic 2-type* of M is the quadruple $[\pi, \pi_2(M), k_1(M), S(\tilde{M})]$, where $S(\tilde{M})$ is the intersection form on $\pi_2(M) = H_2(\tilde{M}; Z)$. Two such quadruples $[\pi, \Pi, \kappa, S]$ and $[\pi', \Pi', \kappa', S']$ (corresponding to M and M', respectively) are equivalent if there is an isomorphism $\alpha : \pi \to \pi'$ and an (anti)isometry $\beta : (\Pi, S) \to (\Pi', (\pm)S')$ which is α-equivariant (i.e., such that $\beta(gm) = \alpha(g)\beta(m)$ for all $g \in \pi$ and $m \in \Pi$) and $\beta_*\kappa = \alpha^*\kappa'$ in $H^3(\pi, \alpha^*\Pi')$. (Note that we then have $w_1(M')\circ\alpha = w_1(M)$ and $w_2(\tilde{M}')\circ\beta = w_2(\tilde{M})$, since $\pi_2(M)$ has nonzero rank). If M is orientable and π is cyclic then the equivalence class of the quadratic 2-type determines the homotopy type [HK88]. In particular, if $\pi = 1$ the manifold M is homotopy equivalent to $S^2 \times S^2$, $S^2 \tilde{\times} S^2$ or $CP^2 \sharp CP^2$.

The argument of [HK88] has been reformulated and extended in Theorem II.7.12 of [Ba']. If M is nonorientable we may adapt Baues' argument by using the orientation π-module $\tilde{Z}$ and the w_1-twisted transfer $\tilde{tr} : \Gamma(\pi_2) \otimes_\pi \tilde{Z} \to \Gamma(\pi_2)$, defined by $\tilde{tr}(x \otimes 1) = \Sigma_{g\in\pi} w_1(g)x.g$, instead of the augmentation module Z and the ordinary transfer. In all cases $\Gamma(Z^2) \cong \Gamma(Z) \oplus \Gamma(Z) \oplus (Z \odot Z) \cong Z^3$, as an abelian group.

The two inclusions of S^2 as factors of $S^2 \times S^2$ determine the standard basis for $\pi_2(S^2 \times S^2)$. Let $J = \left(\begin{smallmatrix} 0 & 1 \\ 1 & 0 \end{smallmatrix}\right)$ be the matrix of the intersection form $\bullet$ on $\pi_2(S^2 \times S^2)$, with respect to this basis. The group $Aut(\pm\bullet)$ of automorphisms of $\pi_2(S^2 \times S^2)$ which preserve this intersection form up to sign is the dihedral group of order eight, and is generated by the diagonal matrices and J or $K = \left(\begin{smallmatrix} 0 & 1 \\ -1 & 0 \end{smallmatrix}\right)$. The subgroup of strict isometries has order four, and is generated by $-I$ and J. (Note that the isometry J is induced by the involution τ).

Let f be a self homeomorphism of $S^2 \times S^2$ and let f_* be the induced automorphism of $\pi_2(S^2 \times S^2)$. The Lefshetz number of f is $2 + trace(f_*)$ if f is orientation preserving and $trace(f_*)$ if f is orientation reversing. As any self homotopy equivalence which induces the identity on π_2 has nonzero Lefshetz number the natural representation of a group π of fixed point free self homeomorphisms of $S^2 \times S^2$ into $Aut(\pm\bullet)$ is faithful.

Suppose first that f is a free involution, so $f_*^2 = I$. If f is orientation preserving then $trace(f_*) = -2$ so $f_* = -I$. If f is orientation reversing then $trace(f_*) = 0$, so $f_* = \pm JK = \pm\left(\begin{smallmatrix} 1 & 0 \\ 0 & -1 \end{smallmatrix}\right)$. Note that if $f' = \tau f \tau$ then $f'_* = -f_*$, so after conjugation by τ if necessary we may assume that $f_* = JK$.

If f generates a free $Z/4Z$-action the induced automorphism must be $\pm K$. Note that if $f' = \tau f \tau$ then $f'_* = -f_*$, so after conjugation by τ if necessary we may assume that $f_* = K$.

Since the orbit space of a fixed point free action of $(Z/2Z)^2$ on $S^2 \times S^2$ has Euler characteristic 1 it is nonorientable, and so the action is generated by two commuting involutions, one of which is orientation preserving and one of which is not. Since the orientation preserving involution must act via $-I$ and the orientation reversing involutions must act via $\pm JK$ the action of $(Z/2Z)^2$ is essentially unique.

The standard inclusions of $S^2 = CP^1$ into the summands of $CP^2\sharp -CP^2 \cong S^2\tilde{\times}S^2$ determine a basis for $\pi_2(S^2\tilde{\times}S^2) \cong Z^2$. Let $\tilde{J} = \left(\begin{smallmatrix} 1 & 0 \\ 0 & -1 \end{smallmatrix}\right)$ be the matrix of the intersection form $\tilde{\bullet}$ on $\pi_2(S^2\tilde{\times}S^2)$ with respect to this basis. The group $Aut(\pm\tilde{\bullet})$ of automorphisms of $\pi_2(S^2\tilde{\times}S^2)$ which preserve this intersection form up to sign is the dihedral group of order eight, and is also generated by the diagonal matrices and $J = \left(\begin{smallmatrix} 0 & 1 \\ 1 & 0 \end{smallmatrix}\right)$. The subgroup of strict isometries has order four, and consists of the diagonal matrices. A nontrivial group of fixed point free self homeomorphisms of $S^2\tilde{\times}S^2$ must have order 2, since $S^2\tilde{\times}S^2$ admits no fixed point free orientation preserving involution. If f is an orientation reversing free involution of $S^2\tilde{\times}S^2$ then $f_* = \pm J$. The automorphism induced by the fibrewise antipodal map is $-J$, as this map clearly changes the sign of the homotopy class of the fibre, which has self intersection 0. Is J also realizable by an orientation reversing free involution?

It is easily seen that all self homeomorphisms of $CP^2\sharp CP^2$ preserve the sign of the intersection form, and thus are orientation preserving. With Lemma 3(ii), this implies that no manifold in this homotopy type admits a free involution.

5. The homotopy type - k-invariants.

In this section we shall try to determine the possible homotopy types. Our results depend in part on [KKR 92] and [HKT 92], and are incomplete when $\pi \cong (Z/2Z)^2$.

Lemma 4. *Let M be a closed 4-manifold with fundamental group $\pi = Z/2Z$ and universal covering space $S^2 \times S^2$. Then the first k-invariant of M is a nonzero element of $H^3(\pi; \pi_2(M))$.*

Proof. The first k-invariant is the primary obstruction to the existence of a cross-section to the classifying map $c_M : M \to K(Z/2Z, 1) = RP^\infty$ and is the only obstruction to the existence of such a cross-section for $g_M : M \to P_2(M)$. The only nonzero differentials in the Cartan-Leray cohomology spectral sequence (with coefficients $Z/2Z$) for the projection $p : \tilde{M} \to M$ are at the E_3^{**} level. By the results of Section 4, π acts trivially on $H^2(\tilde{M}; Z/2Z)$, since $\tilde{M} = S^2 \times S^2$. Therefore $E_3^{22} = E_2^{22} \cong (Z/2Z)^2$ and $E_3^{50} = E_2^{50} = Z/2Z$. Hence $E_\infty^{22} \neq 0$, so E_∞^{22} maps onto $H^4(M; Z/2Z) = Z/2Z$ and $d_3^{12} : H^1(\pi; H^2(\tilde{M}; Z/2Z)) \to H^4(\pi; Z/2Z)$ must be onto. But in this region the spectral sequence is identical with the corresponding spectral sequence for $P_2(M)$. It follows that the image of $H^4(\pi; Z/2Z) = Z/2Z$ in $H^4(P_2(M); Z/2Z)$ is 0, and so g_M does not admit a cross-section. Thus $k_1(M) \neq 0$. //

If $\pi = Z/2Z$ and M is orientable then π acts via $-I$ on Z^2 and the k-invariant is a nonzero element of $H^3(Z/2Z; \pi_2(M)) = (Z/2Z)^2$. The isometry which transposes the standard generators of Z^2 is π-linear, and so there are just two equivalence classes of quadratic 2-types to consider. The k-invariant which is invariant under transposition is realised by $(S^2 \times S^2)/(-I, -I)$, while the other k-invariant is realized by the orientable bundle space with $w_2 = 0$ [HK88]. Thus M must be homotopy equivalent to one of these spaces.

If $\pi = Z/2Z$, M is nonorientable and $w_2(\tilde{M}) = 0$ then $H^3(\pi; \pi_2(M)) = Z/2Z$ and there is only one quadratic 2-type to consider. As $\mathrm{Ker}\tilde{tr} \cong (Z/2Z)^2$ there are at most four homotopy types of 4-dimensional Poincaré complexes within this quadratic 2-type. The product space $S^2 \times RP^2$ is characterized by the additional conditions that $w_2(M) = w_1(M)^2 \neq 0$ (i.e., the Wu class $v_2(M) = w_2(M) + w_1(M)^2$ is 0) and that there is an element $u \in H^2(M;Z)$ which generates an infinite cyclic direct summand and is such that $u \cup u = 0$. (See Theorem IV.7). The nontrivial nonorientable S^2-bundle over RP^2 has $w_2(M) = 0$. The third homotopy type is represented by $RP^4 \sharp_{S^1} RP^4$, which is also the union of two disc bundles over RP^2 and has $w_2(M) = 0$. However it may be distinguished from the nontrivial bundle space by the quadratic function $q : \pi_2(M) \otimes (Z/2Z) \to Z/4Z$ introduced in [KKR92], and is not geometric. There is a fourth homotopy type which is not realizable by a closed manifold [HM78, KKR92].

If $\pi = Z/2Z$ and $w_2(\tilde{M}) \neq 0$ then there are two possible actions of π on Z^2; we do not know whether both can be realized, but in either case $H^3(\pi_1; \pi_2(M)) = 0$. (Note that in this case the argument of Lemma 4 breaks down because $E_\infty^{22} = 0$). The nontrivial RP^2-bundle over S^2 is characterized by the additional condition that there is an element $u \in H^2(M;Z)$ which generates an infinite cyclic direct summand and such that $u \cup u = 0$. (See Theorem IV.7). As $\mathrm{Ker}\tilde{tr} = Z/2Z$ there is at most one other homotopy type with the same quadratic 2-type as this bundle space. In [HKT92] it is shown that any closed 4-manifold M with $\pi = Z/2Z$, $\chi(M) = 2$ and $w_2(\tilde{M}) \neq 0$ is homotopy equivalent to the bundle space.

If $\pi \cong Z/4Z$ then $H^3(\pi; \pi_2(M)) \cong \mathrm{Ker}(I + f_* + f_*^2 + f_*^3)/(I - f_*) = Z^2/(I - K)Z^2 = Z/2Z$. The k-invariant is nonzero, since it restricts to the k-invariant of the orientation double cover. In this case $\tilde{tr}$ is injective and so M is homotopy equivalent to $(S^2 \times S^2)/\tau(I, -I)$.

Suppose now that $\pi \cong (Z/2Z)^2$ is the diagonal subgroup of $Aut(\pm\bullet) < GL(2, Z)$, and let α be the automorphism induced by conjugation by J. The standard generators of $\pi_2(M) = Z^2$ generate complementary π-submodules, so that $\pi_2(M)$ is the direct sum $\tilde{Z} \oplus \alpha^*\tilde{Z}$ of two infinite cyclic modules. The isometry $\beta = J$ which transposes the factors is α-equivariant, and π and $V = \{\pm I\}$ act nontrivially on each summand. If ρ is the kernel of the action of π on $\tilde{Z}$ then $\alpha(\rho)$ is the kernel of the action on $\alpha^*\tilde{Z}$, and $\rho \cap \alpha(\rho) = 1$. As the projection of $\pi = \rho \oplus V$ onto V is compatible with the action, $H^*(V; \tilde{Z})$ is a direct summand of $H^*(\pi; \tilde{Z})$. This implies in particular that the differentials in the LHSSS $H^p(V;H^q(\rho;\tilde{Z}))\Rightarrow H^{p+q}(\pi;\tilde{Z})$ which end on the row $q=0$ are all 0. Hence $H^3(\pi_1;\tilde{Z})\cong H^1(V;Z/2Z)\oplus H^3(V;\tilde{Z})\cong (Z/2Z)^2$. Similarly $H^3(\pi_1;\alpha^*\tilde{Z})\cong (Z/2Z)^2$, and so $H^3(\pi_1;\pi_2(M))\cong (Z/2Z)^4$. The k-invariant must restrict to the k-invariant of each double cover, which must be nonzero, by Lemma 4. As $H^3(\rho;\tilde{Z})=H^3(\alpha(\rho);\alpha^*\tilde{Z})=0$ and $H^3(\rho;\alpha^*\tilde{Z})=H^3(\alpha(\rho);\tilde{Z})=Z/2Z$, there are at most four possible k-invariants. Moreover the automorphism α and the isometry $\beta=J$ act on the k-invariant by transposing the factors. As there are either 0 or two such k-invariants which are invariant under this transposition, there are at most three equivalence classes of quadratic 2-types to be considered. In this case $\mathrm{Ker}\tilde{tr}\cong (Z/2Z)^2$, and so there are at most 12 homotopy types of such manifolds.

6. Surgery

In the present context every homotopy equivalence is simple since $Wh(\pi)=0$ for all groups π of order ≤ 4 [Hg40].

Suppose first that $\pi=Z/2Z$. Then $H^1(M;Z/2Z)=Z/2Z$ and $\chi(M)=2$, so $H^2(M;Z/2Z)\cong (Z/2Z)^2$. Hence if M is orientable $[M;G/TOP]\cong Z\oplus (Z/2Z)^2$. The surgery obstruction groups are $L_5(Z/2Z,+)=0$ and $L_4(Z/2Z,+)\cong Z^2$, where the surgery obstructions are determined by the signature and the signature of the double cover, by Theorem 13.A.1 of [W]. If M is nonorientable then $[M;G/TOP]\cong (Z/2Z)^3$, and the surgery obstruction of a 4-dimensional normal map $g:M\to G/TOP$ is the Kervaire-Arf invariant $c(g)\in L_4(Z/2Z,-)=Z/2Z$, while $L_5(Z/2Z,-)=0$, by Theorem 13.A.1 of [W]. Now $c(g)=w(M)g^*\kappa[M]=(w_2(M)\cup g^*k_2+g^*Sq^2k_2)[M]=((w_2(M)+v_2(M))\cup g^*k_2)[M]=(w_1(M)^2\cup g^*k_2)[M]$, by Theorem 13.B.5 of [W]. As $w_1(M)$ is not the reduction of a class in $H^1(M;Z/4Z)$ its square $w_1(M)^2$ is nonzero; as every element of $H^2(M;Z/2Z)$ is equal to g^*k_2 for some such g the map $c:[M;G/TOP]\to Z/2Z$ is onto. Hence it follows from the surgery exact sequence that $S_{TOP}(M)\cong (Z/2Z)^2$ whenever $\pi=Z/2Z$.

The $Z/2Z$-Hurewicz homomorphism from $\pi_2(M)$ to $H_2(M;Z/2Z)$ has cokernel $H_2(\pi;Z/2Z)=Z/2Z$. Since $H_2(M;Z/2Z)\cong (Z/2Z)^2$ there is a map $\beta:S^2\to M$ such that $\beta_*[S^2]\neq 0$ in $H_2(M;Z/2Z)$. If moreover $w_2(\tilde{M})=0$ then $\beta^*w_2(M)=0$, since β factors through $\tilde{M}$. Then there is a self homotopy equivalence f_β of M with nontrivial normal invariant in $[M; G/TOP]$, by Lemma V.8. Hence there are at most two homeomorphism classes within the homotopy type of M if $\pi = Z/2Z$ and $w_2(\tilde{M}) = 0$. In [HKT92] it is shown that if M is nonorientable and $\pi = Z/2Z$ then there are two homeomorphism types within each homotopy type if $w_2(\tilde{M}) = 0$; if $w_2(\tilde{M}) \neq 0$ (i.e., if $M \simeq RP^4 \sharp CP^2$) then there are four corresponding homeomorphism types.

When $\pi \cong Z/4Z$ or $(Z/2Z)^2$ the manifold M is nonorientable, since $\chi(M) = 1$. If $\pi \cong Z/4Z$ then $[M; G/TOP] \cong (Z/2Z)^2$ and the surgery obstruction groups $L_4(Z/4Z, -)$ and $L_5(Z/4Z, -)$ are both 0, by Theorem 3.4.5 of [Wl76]. Hence $S_{TOP}(M) \cong (Z/2Z)^2$. If $\pi \cong (Z/2Z)^2$ then $[M; G/TOP] \cong (Z/2Z)^4$ and the surgery obstruction groups are $L_5((Z/2Z)^2, -) = 0$ and $L_4((Z/2Z)^2, -) = Z/2Z$, by Theorem 3.5.1 of [Wl76]. The homomorphism $w_1(M) : (Z/2Z)^2 \to Z/2Z$ induces an isomorphism from $L_4((Z/2Z)^2, -)$ onto $L_4(Z/2Z, -)$ and so the surgery obstruction is again given by the Kervaire-Arf invariant. We now find that $S_{TOP}(M) \cong (Z/2Z)^3$. The argument above for the existence of exotic self homotopy equivalences does not apply as the $Z/2Z$-Hurewicz homomorphism is 0 in these cases.

The image of $[M; G/PL]$ in $[M; G/TOP]$ is a subgroup of index 2 (see Section 15 of [Si71]). It follows that if M is the total space of an S^2-bundle over RP^2 any homotopy equivalence $f : N \to M$ where N is also PL is homotopic to a homeomorphism. (For then $S_{TOP}(M)$ has order 4, and the nontrivial element of the image of $S_{PL}(M)$ is represented by an exotic self homotopy equivalence of M. The case $M = S^2 \times RP^2$ was treated in [Ma79]). This is also true if $M = S^4$, RP^4, CP^2, $S^2 \times S^2$ or $S^2 \tilde{\times} S^2$. Two of the exotic homeomorphism types within the homotopy type of $RP^4 \sharp CP^2$ (the nontrivial RP^2-bundle over S^2) have nontrivial Kirby-Siebenmann invariant. The third has trivial Kirby-Siebenmann invariant and is stably homeomorphic to $RP^4 \sharp CP^2$ (after connected sum with one copy of $S^2 \times S^2$), but it is not known whether it is PL [HKT92]. What is the situation for the remaining three $S^2 \times S^2$-manifolds?

CHAPTER X

APPLICATIONS TO 2-KNOTS AND COMPLEX SURFACES

A 2-knot is a locally flat embedding $K : S^2 \to S^4$. The closed 4-manifold $M(K)$ obtained from S^4 by surgery on K is orientable, has Euler characteristic 0 and $\pi K = \pi_1(M(K))$ has weight 1 (i.e., is the normal closure of a single element) and infinite cyclic abelianization, $\pi K/\pi K' = H_1(M(K); Z) = Z$. Conversely, if M is a closed orientable 4-manifold with $\chi(M) = 0$ and $\pi_1(M)$ of weight 1 then it may be obtained in this way, for surgery on a loop in M representing a normal generator for $\pi_1(M)$ gives a 1-connected 4-manifold Σ with $\chi(\Sigma) = 2$ which is thus homeomorphic to S^4 and which contains an embedded 2-sphere as the cocore of the surgery. If $\pi_1(M)$ is solvable then it has weight 1 if and only if $\pi_1(M)/\pi_1(M)'$ is cyclic, for a solvable group with trivial abelianization must be trivial. (See [H] for more details).

In the next two sections we shall summarize progress made on the study of 2-knot groups since the appearance of [H]. We shall then consider when the manifolds $M(K)$ admit geometries or complex analytic structures. The final section gives new characterizations of minimal complex surfaces which are ruled over curves of genus > 1 or are elliptic surfaces, fibred over such curves.

1. Elementary amenable 2-knot groups

Let K be 2-knot. Then πK is finitely presentable and has infinite cyclic abelianization, and so it is an HNN extension with finitely generated base and associated subgroups, by Theorem A of [BS78]. Suppose that πK is amenable. Then πK has no noncyclic free subgroup, and so the HNN extension must be ascending. Moreover πK has one or two ends. There are three cases:

(i) πK has two ends. (Equivalently, $\pi K'$ is finite).

All such groups are determined in Chapter IV of [H].

Typeset by $\mathcal{A}_{\mathcal{M}}\mathcal{S}$-TeX

(ii) πK has one end and $H^2(\pi K; Z[\pi K]) = 0$.

In this case $M(K)$ is aspherical and so πK is a PD_4-group [Ec93]. Therefore if moreover πK is elementary amenable then it is a torsion free virtually poly-Z group of Hirsch length 4. All such groups are determined in Chapter VI of [H].

(iii) πK has one end and $H^2(\pi K; Z[\pi K]) \neq 0$.

The only known example of such an amenable 2-knot group is Fox's 2-knot group Φ, with presentation $< a, t | tat^{-1} = a^2 >$.

We have two arguments that suggest there may be no other amenable 2-knot groups of type *(iii)*. As the base of the HNN extension is amenable it must have finitely many ends, and as πK has one end the base must be infinite. If the base is finitely *presentable* then it must have two ends, for otherwise $H^2(\pi K; Z[\pi K]) = 0$ by [BG85]. In that case it follows quickly that πK is an extension of Φ by a finite normal subgroup; by Theorem IV.6 of [H] we then have $\pi K \cong \Phi$.

If we knew that $h(G) > 2$ implies $H^s(G; Z[G]) = 0$ for $s \leq 2$ for any *elementary* amenable group G we could conclude that if πK is an elementary amenable 2-knot group of type *(iii)* then $h(\pi K) = 2$, and then $\pi K/T \cong \Phi$, where T is the maximal locally-finite normal subgroup of πK. If T is finite it must be trivial, by Theorem IV.6 of [H]; no example is known with T infinite.

The discussion of Nil^3-manifolds on pages 100-102 of [H] overlooks examples such as circle bundles over the Klein bottle. However there is a purely algebraic argument which implies that if $H = \sqrt{\pi K'} \cong \Gamma_q$ for some $q \geq 1$ then πK must be one of the groups listed in Theorems 10 and 11 on those pages. Since $\zeta H \cong Z$ is normal in πK it is central in $\pi K'$. Using the known structure of automorphisms of Γ_q, it follows that the finite group $\pi K'/H$ must act on $H/\zeta H \cong Z^2$ via $SL(2, Z)$ and so must be cyclic.

It is not obvious that knots with $\pi' \cong \Gamma_1$ and Alexander polynomial $X^2 - 3X + 1$ are +amphicheiral (as asserted on page 131 of [H]), but this is true. (The argument uses the fact that a 2×2 matrix with characteristic polynomial $X^2 - X + 1$ or $X^2 - 3X + 1$ is in a 1-parameter subgroup of $GL(2, R)$, and does not extend to the knots with $\pi' \cong \Gamma_q$ and Alexander polynomial $X^2 - X - 1$).

The question of which Cappell-Shaneson 2-knots are reflexive has been settled in [HW89]. Only the knots with Alexander polynomial $X^3 + X^2 - 2X - 1$ are reflexive. (Up to change of orientation there is just one such 2-knot with metabelian knot group).

2. Abelian normal subgroups

We may summarize what is currently known about maximal abelian normal subgroups A of 2-knot groups πK as follows:
(i) if A has rank 1 and $A \not\subseteq \pi K'$ then $A \cong Z$ or $\pi K'$ is finite; in the latter case the groups πK are well understood. (See Chapter IV of [H] and [Hi93]).
(ii) if A has rank 1 and $A \subseteq \pi K'$ then either A is torsion free or $\pi K'/A$ is not finitely generated. (See Chapter IV of [H]).
(iii) if A has rank 2 and $A \not\subseteq \pi K'$ then A is torsion free, and $A \cong Z^2$ if $\pi K'$ has a subgroup of finite index with infinite abelianization. (See Chapter V of [H] and also [Hi93]).
(iv) if A has rank 2, is torsion free and $A \subseteq \pi K'$ then $A \cong Z^2$ and $\pi K'$ is not finitely generated. (No such example is known. See Chapter V of [H]).
(v) if A has rank greater than 2 then $A \cong Z^3$ or Z^4 and the groups πK are well understood. (See Chapter VI of [H]).
The evidence suggests that if $\pi K'$ is finitely generated and infinite then A is free abelian. (Note however that $\Phi' \cong Z[\frac{1}{2}]$). Very little is known about the rank 0 case, although Yoshikawa has constructed examples with A finite cyclic. He has also shown that if K has a minimal Seifert hypersurface and A is finitely generated then $A \cap \pi K'$ is finite cyclic or is torsion free [Yo92].

The localization strategy extends to the study of 2-knot groups with nontrivial elementary amenable normal subgroups (with restricted torsion). If $h(G) > 2$ implies $H^s(G; W) = 0$ for $s \leq 2$ for any elementary amenable group G and free $Z[G]$-module W then there are no new examples, as elementary amenable groups of finite cohomological dimension are torsion free and virtually solvable. (This connection between Hirsch length and vanishing of cohomology is valid if the group is locally nilpotent, by Theorem I.8).

3. Geometries and 2-knots

We shall apply the criteria of Chapters VI and VII to the question of which 2-knot manifolds $M(K)$ admit a geometry. In the infrasolvmanifold cases the corresponding knots are essentially known. The other realizable geometries are product geometries; all known examples are related to twist spins of classical knots. For one geometry only ($H^2 \times E^2$) we do not know whether there is such a knot manifold.

Theorem 1. *Let K be a 2-knot. Then $M(K)$ is homeomorphic to an infrasolvmanifold if and only if πK has a locally nilpotent normal subgroup of Hirsch length at least 3, by Theorem VI.2. No such knot manifold $M(K)$ admits a geometry of type Nil^4, S^4, CP^2, $S^2 \times S^2$, $S^2 \times E^2$, $S^2 \times H^2$, H^4, $H^2(C)$, $H^2 \times H^2$ or F^4.*

Proof. If $M(K)$ is an infrasolvmanifold then the above condition on πK is certainly necessary. Suppose that it holds. Then πK determines $M(K)$ up to homeomorphism, by Theorem V.7. By Theorem VI.14 of [H] πK must be either $G(+)$ or $G(-)$, $\pi(b,\epsilon)$ for some even b and $\epsilon = \pm 1$ or $\pi K' \cong Z^3$ or Γ_q for some odd q.

If $\pi K \cong G(+)$ or $G(-)$ then it is virtually Z^4 and so $M(K)$ admits a geometry of type E^4. If $\pi K \cong \pi(b,\epsilon)$ then $M(K)$ is the mapping torus of the canonical involution of the 2-fold branched cover of S^3, branched over the Montesinos knot $K(o|b;(3,1),(3,1),(3,\epsilon))$, which is a Nil^3-manifold, and so $M(K)$ admits a geometry of type $Nil^3 \times E^1$. If $\pi K' \cong Z^3$ then K is a Cappell-Shaneson 2-knot, and up to change of orientations we may assume that the Alexander polynomial of K is $X^3 - mX^2 + (m-1)X - 1$ for some integer m. If $m \geq 6$ all the roots of this cubic are positive and $M(K)$ admits a geometry of type $Sol^4_{m,m-1}$. If $0 \leq m \leq 5$ two of the roots are complex conjugates and $M(K)$ admits a geometry of type Sol^4_0. If $m < 0$ two of the roots are negative and πK has a subgroup of index 2 which is a discrete cocompact subgroup of the Lie group $Sol^4_{m',n'}$, where $m' = m^2 - 2m + 2$ and $n' = m^2 - 4m + 1$, so $M(K)$ admits a geometry of type $Sol^4_{m',n'}$. If $\pi K' \cong \Gamma_q$ and the meridianal outer automorphism is of finite order then $q = 1$ and K is the 6-twist spin of the trefoil knot or its 5-fold branched cover (which is also its Gluck reconstruction). In this case $M(K)$ admits a geometry of type $Nil^3 \times E^1$. Otherwise (if $\pi K' \cong \Gamma_q$ and the meridianal outer automorphism is of infinite order) $M(K)$ admits a geometry of type Sol_1^4.

It follows from the above discussion that there is no 2-knot K for which $M(K)$ admits a geometry of type Nil^4 and that many of the $Sol_{m,n}^4$ geometries do not occur. (In particular, $Sol^3 \times E^1$ does not arise in this way). The geometries S^4, $P^2(C)$ and $S^2 \times S^2$ cannot arise since πK is infinite; the geometries H^4, $H^2(C)$, $H^2 \times H^2$ and $S^2 \times H^2$ cannot arise since $\chi(M(K)) = 0$ (see [Wl86] and [Ko92]); the geometry $S^2 \times E^2$ cannot arise since no group which is virtually Z^2 can have infinite cyclic abelianization (see Theorem VII.5) and the geometry F^4 is not realized by any closed 4-manifold. //

If πK has a torsion free elementary amenable normal subgroup ρ with $h(\rho) = 2$ then by Theorem VI.11 either ρ has finite index in πK, in which case $\pi K \cong \Phi$, by Theorem IV.6 of [H], or ρ is virtually abelian. In the latter case πK has a torsion free abelian normal subgroup of rank 2 and $M(K)$ is aspherical. Less is known about the cases when $h(\rho) = 1$ (and so ρ is abelian of rank 1) and there is no such subgroup of greater length.

We do not know whether there are any 2-knots K such that $M(K)$ admits a geometry of type $H^2 \times E^2$. If there is such a knot then $\sqrt{\pi K} \cong Z^2$ and $G = \pi K/\sqrt{\pi K}$ is virtually a surface group. If $\sqrt{\pi K} \leq \pi K'$ then G/G' is infinite cyclic and so G has a finite normal subgroup N such that G/N is a plane motion group [EM82]. but examination of the presentations of such groups, as given in Theorem 4.5.6 of [Z], shows that no such group has infinite cyclic abelianization. Therefore $\sqrt{\pi K}$ is not contained in $\pi K'$ and $M(K)$ must fibre over S^1 with fibre an $H^2 \times E^1$-manifold and monodromy of finite order.

The manifolds obtained from branched r-twist spins of (p, q)-torus knots with $p^{-1} + q^{-1} + r^{-1} < 1$ have geometry $\tilde{SL} \times E^1$, while those obtained from branched r-twist spins of simple (nontorus) knots with $r > 2$ have geometry $H^3 \times E^1$, excepting only the one from the 3-twist spin of the figure eight knot and its 2-fold branched cover (which is also its Gluck reconstruction), which has group $G(+)$ and geometry E^4. (The latter manifold is the only closed orientable E^4-manifold which is not "Seifert fibred" [Ue90]). Examples with geometry $S^3 \times E^1$ arise from 2-twist spins of 2-bridge knots and certain other "small" simple knots, and from branched r-twist spins of (p, q)-torus knots with $p^{-1}+q^{-1}+r^{-1}>1$. (The case $p^{-1}+q^{-1}+r^{-1}=1$ gives a $Nil^3\times E^1$-manifold).

4. Complex surfaces and 2-knots

In what follows *complex surface* shall mean compact connected nonsingular complex analytic manifold of complex dimension 2. If $M(K)$ is homeomorphic to a complex surface S then S is minimal, since $\beta_2(S)=0$, and has Kodaira dimension $\kappa(S)=1$, 0 or -1, since $\beta_1(S)=1$ is odd. If $\kappa(S)=1$ or 0 then S is elliptic and admits a compatible geometric structure, of type $\tilde{SL}\times E^1$ or $Nil^3\times E^1$, respectively [Ue90,91, Wl86]. The only complex surfaces with $\kappa(S)=-1$, $\beta_1(S)=1$ and $\beta_2(S)=0$ are Inoue surfaces, which are not elliptic, but admit compatible geometries of type Sol_0^4 or Sol_1^4, and Hopf surfaces [LYZ90]. A Hopf surface is a complex surface whose universal covering space is homeomorphic to $S^3\times R\cong C^2\backslash\{0\}$. Some Hopf surfaces admit no compatible geometry [Wl85], and there are $S^3\times E^1$-manifolds that admit no complex structure. The geometric Hopf surfaces are the elliptic surfaces of Kodaira dimension -1. By Theorem 4.5 of [Wl86], if $M(K)$ has a complex structure compatible with a geometry then the geometry is one of Sol_0^4, Sol_1^4, $Nil^3\times E^1$, $S^3\times E^1$ or $\tilde{SL}\times E^1$. Conversely, if $M(K)$ admits one of the first three of these geometries then it is homeomorphic to an Inoue surface of type S_M, an Inoue surface of type $S^{(+)}_{N,p,q,r;t}$ or $S^{(-)}_{N.p,q,r}$ or an elliptic surface of Kodaira dimension 0, respectively. (See [In74] and Chapter V of [BPV]).

Lemma 2. *Let K be a branched r-twist spin of the (p,q)-torus knot. Then $M(K)$ is homeomorphic to an elliptic surface.*

Proof. We shall adapt the argument of Lemma 1.1 of [Mi75]. (See also [Ne83]). Let $V_0=\{(z_1,z_2,z_3)\in C^3\backslash\{0\}|z_1^p+z_2^q+z_3^r=0\}$, and define an action of $C^\times$ on V_0 by $u.v=(u^{qr}z_1,u^{pr}z_2,u^{pq}z_3)$ for all u in $C^\times$ and $v=(z_1,z_2,z_3)$ in V_0. Define functions $m:V_0\to R^+$ and $n:V_0\to m^{-1}(1)$ by $m(v)=(|z_1|^p+|z_2|^q+|z_3|^r)^{1/pqr}$ and $n(v)=m(v)^{-1}.v$ for all v in V_0. Then the map $(m,n):V_0\to m^{-1}(1)\times R^+$ is an R^+-equivariant homeomorphism, and so $m^{-1}(1)$ is homeomorphic to V_0/R^+. Therefore there is a homeomorphism from $m^{-1}(1)$ to the Brieskorn manifold $M(p,q,r)$, under which the action of the group of rth roots of unity on $m^{-1}(1) = V_0/R^+$ corresponds to the group of covering homeomorphisms of $M(p,q,r)$ as the branched cyclic cover of S^3, branched over the (p,q)-torus knot [Mi75]. The manifold $M(K)$ is the mapping torus of some generator of this group of self homeomorphisms of $M(p,q,r)$. Let ω be the corresponding primitive r^{th} root of unity. If $t > 1$ then $t\omega$ generates a subgroup Ω of $C^\times$ which acts freely and holomorphically on V_0, and the quotient V_0/Ω is an elliptic surface over the curve V_0/Ω. Moreover V_0/Ω is homeomorphic to the mapping torus of the self homeomorphism of $m^{-1}(1)$ which maps v to $m(t\omega.v)^{-1}.t\omega.v = \omega m(t.v)^{-1}t.v$ Since this map is isotopic to the map sending v to $\omega.v$ this mapping torus is homeomorphic to $M(K)$. This proves the Lemma. //

The Kodaira dimension of the elliptic surface in the above lemma is 1, 0 or -1 according as $p^{-1}+q^{-1}+r^{-1}$ is < 1, 1 or > 1. In the next theorem we shall settle the case of elliptic surfaces with $\kappa = -1$.

Theorem 3. *Let K be a 2-knot. Then $M(K)$ is homeomorphic to a Hopf surface if and only if K is a branched r-twist spin of the (p,q)-torus knot for some p, q and r such that $p^{-1}+q^{-1}+r^{-1} > 1$.*

Proof. If K is such a branched twist spin then $M(K)$ is homeomorphic to an elliptic surface, by the lemma, and the surface must be a Hopf surface if $p^{-1}+q^{-1}+r^{-1} > 1$.

If $M(K)$ is homeomorphic to a Hopf surface then either K is trivial or πK is nonabelian; in either case πK is isomorphic to a subgroup of $GL(2,C)$ which contains a contraction (Kodaira - see [Ka75]). since the image of πK under the homomorphism $det : GL(2,C) \to C^\times$ is infinite and abelian $\pi K \cap SL(2,C) = \{g \in \pi K \| det(g)| = 1\} = \pi K'$. It then follows from Proposition 8 of [Ka75] and Theorem IV.3 of [H] that $\pi K'$ is either cyclic of odd order n, the quaternion group (B_2 in the notation of [Ka75]), binary tetrahedral or binary icosahedral. (There is an error in Lemma 4 of [Ka75] - the normalizer $N_{SL(2,C)}(B_2)$ has an element of order 3). Hence πK is isomorphic to the group of the 2-twist spin of the $(2,n)$-torus knot, or of the 3-, 4- or 5-twist spin of the trefoil knot, respectively. Since such Hopf surfaces are determined up to diffeomorphism by their fundamental groups, by Theorems 10 and 12 of [Ka75], $M(K)$ is homeomorphic to the manifold of the corresponding torus knot. Since each weight class in one of these groups is realized by some branched twist spin of such a torus knot (see Table 1 of [PS87]) and the Gluck reconstruction of a branched r-twist spin of a classical knot is another branched r-twist spin of that knot [Pl84] the theorem follows. //

As observed above, a knot manifold $M(K)$ is homeomorphic to an elliptic surface with $\kappa = 0$ if and only if it admits a geometry of type $Nil^3 \times E^1$. Such knots may be characterized algebraically by the conditions πK is virtually poly-Z and $\zeta\pi K \cong Z^2$. For the remaining class of elliptic surfaces (those with $\kappa = 1$) we must settle for a characterization up to simple homotopy equivalence.

Theorem 4. *Let K be a 2-knot. Then $M(K)$ is simple homotopy equivalent to an elliptic surface of Kodaira dimension 1 if and only if $\zeta\pi K \cong Z^2$ and $\pi K'$ has a subgroup of finite index with infinite abelianization but is neither virtually poly-Z nor virtually a product $Z \times \sigma$.*

Proof. If $M(K)$ is an elliptic surface of Kodaira dimension 1 then it admits a compatible geometry of type $\tilde{SL} \times E^1$ and πK is isomorphic to a discrete cocompact subgroup of $Isom_o(\tilde{SL}) \times R$, the maximal connected subgroup of $Isom_o(\tilde{SL} \times E^1)$, for the other components consist of orientation reversing or antiholomorphic isometries (see Theorem 3.3 of [Wl86]). Since πK meets $\zeta(Isom_o(\tilde{SL}) \times R)) \cong R^2$ in a lattice subgroup $\zeta\pi K \cong Z^2$ and since πK projects nontrivially onto the second factor $\pi K' = \pi K \cap Isom_o(\tilde{SL})$ and is the fundamental group of an $\tilde{SL}$-manifold. Thus the conditions are necessary.

Suppose that they hold. Then $M(K)$ is aspherical, $\pi K'$ is the fundamental group of an aspherical Seifert fibred 3-manifold N and $\zeta\pi K$ is not contained in $\pi K'$, by Theorems 1, 2 and 6 of Chapter 5 of [H]. Since $\pi K'$ is neither virtually poly-Z nor virtually a product N must be of type $\tilde{SL}$. Since $\zeta\pi K$ is not contained in $\pi K'$ the meridianal outer automorphism has finite order. This outer automorphism can be realized by a self homeomorphism Θ of N and the mapping torus $M(\Theta)$ has the product geometry, by Theorem VI.10. As moeover the meridianal automorphism is orientation preserving and induces the identity on $\zeta\pi K' \cong Z$ the group πK is contained in $Isom_o(\tilde{SL})\times R$, and so $M(\Theta)$ has a compatible structure as an elliptic surface of Kodaira dimension 1, by Theorem 3.3 of [Wl86]. Since $M(K)$ is aspherical it is homotopy equivalent to $M(\Theta)$. The torus $\zeta(Isom_o(\tilde{SL})\times R)/\zeta\pi K \cong R^2/Z^2$ acts effectively on $M(\Theta)$. Therefore $Wh(\pi K)=0$ [NS85] and so any such homotopy equivalence is simple.//

The argument of [NS85] also shows that the surgery obstruction maps are bijections. In particular, $L_5(\pi K)\cong Z\oplus(Z/2Z)$. Using Lemma V.12 we may conclude that $S_{TOP}^s(M(\Theta))$ has order at most 2.

An elliptic surface with Euler characteristic 0 is a Seifert fibred 4-manifold, and so is determined up to diffeomorphism by its fundamental group if the base orbifold is euclidean or hyperbolic [Ue90,91]. Using this result (instead of [Ka75]) together with Theorem V.8 of [H] and Lemma 2 above it may be shown that if $M(K)$ is homeomorphic to an elliptic surface of Kodaira dimension 1 and some power of a weight element is central in πK then $M(K)$ is homeomorphic to $M(K_1)$, where K_1 is some branched twist spin of a torus knot. However in general there may be infinitely many algebraically distinct weight classes in πK and we cannot conclude that K is itself such a branched twist spin.

5. Elliptic surfaces and ruled surfaces

We conclude by sketching some applications of results from Chapters II, IV and VI to the characterization of certain complex surfaces.

Theorem 5. *Let S be a complex surface with $\chi(S)=0$ and suppose that $\pi=\pi_1(S)$ has a normal subgroup $A\cong Z^2$ such that π/A is torsion free and has a subgroup of finite index with infinite abelianization.*
(i) If π/A is virtually Z^2 then S is either an Inoue surface, a hyperelliptic surface, a Kodaira surface, a complex torus or is a minimal properly elliptic surface;
(ii) if π/A is not virtually Z^2 then S is a minimal properly elliptic surface.
Proof. The space S is aspherical and the quotient π/A is a PD_2-group, by Theorem VI.12. In particular, S is minimal since $\pi_2(S)=0$. We may assume

that S is a projective algebraic surface, for otherwise (i) and (ii) follow from the Enriques-Kodaira classification of complex surfaces. (See Chapter VI of [BPV], together with [LYZ90]). This classification also gives (i), for if π/A is virtually Z^2 then π is virtually poly-Z.

Therefore we may also assume that π/A is not virtually Z^2, and so it is isomorphic to a discrete torsion free group of isometries of the upper half plane H. Since there are nonzero L^2 harmonic 1-forms on H the ℓ_2-cohomology group $H^1EL_2(\pi/A)$ is nontrivial and so there is a properly discontinuous holomorphic action of π/A on H and a π/A-equivariant holomorphic map from the covering space S_A to H, with connected fibres, by Theorems 4.1 and 4.2 of [ABR92]. The quotient map from H to the complex curve $B = H/(\pi/A)$ is a covering projection, since π/A is torsion free. Hence the induced map $h : S \to B$ induces an epimorphism with kernel A.

The map h is a submersion away from the preimage of a finite subset $D \subset B$. Let F be the general fibre and F_d the fibre over $d \in D$. Then $\chi(S) = \chi(F)\chi(B) + \Sigma_{d\in D}(\chi(F_d) - \chi(F))$ and $\chi(F_d) \geq \chi(F)$, by Proposition III.11.4 of [BPV]. Moreover $\chi(F_d) > \chi(F)$ unless $\chi(F_d) = \chi(F) = 0$, by Remark III.11.5 of [BPV]. Since $\chi(S) = 0$ and $\chi(B) = \chi(\pi) < 0$ we must have $\chi(F) \geq 0$. If $\chi(F) > 0$ then $F \cong CP^1$ and the singular set D must be empty. But then h is a submersion and $\pi_2(S) \cong \pi_2(F) \neq 0$. Therefore as S is aspherical we must have $\chi(F) = 0$, and so S is an elliptic surface. //

We may apply the same strategy to obtain our final theorem.

Theorem 6. *Let S be a complex surface. Then the following are equivalent:*
(i) S is minimal and is ruled over a complex curve of genus > 1;
(ii) $\pi = \pi_1(S)$ is a PD_2-group, $w_1(\pi) = 0$ and $\chi(S) = 2\chi(\pi) < 0$; and
(iii) $\pi_2(S) \cong Z$, π acts trivially on $\pi_2(S)$ and $\chi(S) < 0$.

Proof. Clearly (i) implies (ii), and (ii) implies (iii) by Theorem IV.3. (Note that π acts trivially on $\pi_2(S)$ since both π and S are orientable). Suppose that (iii) holds. Since S is orientable and $\chi(S) \leq 0$ the abelianization $\pi/\pi' = H_1(S; Z)$ is infinite. Therefore π is a PD_2-group and $\chi(S) = 2\chi(\pi)$, by Theorem II.11. Moreover $w_1(\pi) = 0$, as S is orientable and π acts trivially on $\pi_2(S)$. Hence $\beta_1(S) = \beta_1(\pi)$ is even and so S is a Kähler surface. (See Proposition 4.3 of [Wl86]).

As $\chi(\pi) < 0$ the group π is of hyperbolic type, and so $H^1EL_2(\pi) \neq 0$. Therefore, as before, there is a holomorphic map $h : S \to B$ which has connected fibres and which now induces an isomorphism of fundamental groups. The map h is a submersion away from the preimage of a finite subset $D \subset B$. Let F be the general fibre and F_d the fibre over $d \in D$. Since $\chi(S) = 2\chi(\pi) = 2\chi(B)$, $\chi(F) \leq 2$ and $\chi(B) < 0$ The equation $\chi(S) = \chi(F)\chi(B) + \Sigma_{d\in D}(\chi(F_d) - \chi(F))$ now implies that $\chi(F) = 2$, i.e., the general fibre is isomorphic to CP^1, and that $\chi(F_d) = \chi(F)$, for all $d \in D$. Hence D is empty and so S is ruled over the curve B. As $\pi_2(S)$ has no nontrivial free $Z[\pi]$-module summand S is minimal. //

This theorem also follows from the Enriques-Kodaira classification, since the only complex surfaces with negative Euler characteristic are the ruled surfaces.

If $\pi_2(S) \cong Z$ and $\chi(S) = 0$ then π is virtually Z^2. The finite covering space with fundamental group Z^2 is Kähler, and therefore so is S [Wl86]. Since $\beta_1(S) > 0$ and is even, we must have $\pi \cong Z^2$. It then follows from the Enriques-Kodaira classification that S is either ruled over an elliptic curve or is a minimal properly elliptic surface. In the latter case the base of the elliptic fibration is CP^1, there are no singular fibres and there are at most 3 multiple fibres. (See [Ue91]). Thus S may be obtained from a cartesian product $CP^1 \times E$ by logarithmic transformations. (See §V.13 of [BPV]). Must S in fact be ruled?

If $\pi_2(S) \cong Z$ and $\chi(S) > 0$ then $H^1(S; Z) = 0$ (for otherwise π is virtually PD_2 and so $\chi(S) \leq 0$). In particular, S is Kähler and hence projective algebraic. Must S be CP^2?

APPENDIX

3-Dimensional Poincaré duality complexes

The main reason for studying PD-complexes is that they represent the homotopy theory of manifolds. However they also arise in situations where the geometry does not immediately provide a corresponding manifold. For instance, under suitable finiteness assumptions an infinite cyclic covering space of a closed 4-manifold with Euler characteristic 0 will be a PD_3-complex, as we saw in Chapter III. (In other dimensions we could appeal to the Farrell or Stallings fibration theorems at this point). In this appendix we shall summarize briefly what is known about the homotopy types of PD_3-complexes.

In lower dimensions the classification is complete. It is easy to see that the circle is the only PD_1-complex. The 2-dimensional case is already quite difficult, but has been settled by Eckmann, Linnell and Müller, who showed that every PD_2-group is a surface group and hence that every PD_2-complex is homotopy equivalent to a closed surface [EM80, EL81]. (See also Chapter VI of [DD]). In particular, the only PD_2-complexes which fibre properly are the torus and Klein bottle.

There are PD_3-complexes with finite fundamental group which are not homotopy equivalent to any closed 3-manifold. However it is not known whether every PD_3-complex with torsion free fundamental group is homotopy equivalent to a closed 3-manifold.

The fundamental triple of a PD_3-complex P is the triple (π, w, μ) where $\pi = \pi_1(P)$, $w = w_1(P)$ and $\mu = c_{P*}[P] \in H_3(\pi; Z^w)$ is the image of the (twisted) orientation class of P under the classifying map $c_P : P \to K(\pi, 1)$. Hendriks has shown that the fundamental triple is a complete invariant for PD_3-complexes.

Typeset by $\mathcal{A}\mathcal{M}\mathcal{S}$-TEX

Theorem 1. [He] *Two PD_3-complexes are homotopy equivalent if and only if their fundamental triples are isomorphic.* //

Turaev has given a characterization of the possible triples corresponding to a given finitely presentable group and orientation character. Although it is difficult in practice to use it to decide whether a given group can be the fundamental group of a PD_3-complex, he has used this result to deduce a basic splitting theorem.

Theorem 2. [Tu90] *A PD_3-complex is irreducible with respect to connected sum if and only if its fundamental group is indecomposable with respect to free product.* //

Wall has asked whether every PD_3-complex whose fundamental group has infinitely many ends is a proper connected sum [Wl67]. The fundamental group of a PD_3-group is finitely presentable, and so is the fundamental group of a finite graph of (finitely generated) groups in which each vertex group has at most one end and each edge group is finite, by Theorem VI.6.3 of [DD]. It can be shown that if P is indecomposable and orientable and π has infinitely many ends then at least one of the vertex groups must also be finite [Hi93]. There remains the possibility that, for instance, the free product of two copies of the symmetric group on 3 letters with amalgamation over a subgroup of order 2 may be the fundamental group of an orientable PD_3-complex.

A simple application of Poincaré duality shows that $\pi_2(P) \cong \overline{H^1(\pi; Z[\pi])}$ as a left $Z[\pi]$-module. In Theorem II.13 we saw that if π is torsion free but not free then $\pi_2(P)$ is a projective $Z[\pi]$-module and any two of the conditions "π is FF", "P is homotopy equivalent to a finite complex" and "$\pi_2(P)$ is stably free" imply the third. Similar arguments show that $H^2(P; \pi_2(P)) = 0$. Moreover if π is a nontrivial free group then $\pi_2(P)$ has projective dimension 1 and $H^2(P; \pi_2(P)) \cong Z$. (See [Hi93']). If π is not torsion free then the projective dimension of $\pi_2(P)$ is infinite.

The cohomology group $H^2(P; \pi_2(P))$ arises in studying homotopy classes of self homotopy equivalences of P. Hendriks and Laudenbach show that if N is a P^2-irreducible 3-manifold and $\pi_1(N)$ is virtually free then $H^2(N; \pi_2(N)) \cong Z$, and otherwise $H^2(N;\pi_2(N)) = 0$ [HL74]. Does this result extend to PD_3-complexes with virtually torsion free fundamental group? (If the answer to Wall's question is "yes" then it follows from Turaev's Theorem that all PD_3-complexes have such fundamental groups).

2. The spherical cases

If P is a PD_3-complex with finite fundamental group π then P is orientable, $\tilde{P} \simeq S^3$ and π has cohomological period dividing 4. Conversely, every such group is the fundamental group of some PD_3-complex. The class μ is then a generator of the cyclic group $H_3(\pi;Z) \cong H^4(\pi;Z)$ and is essentially equivalent to the first k-invariant; whether the corresponding complex is homotopy equivalent to a finite complex depends on a coset in $\tilde{K}_0(Z[\pi])$. (See [Wl67], [DM85] and also Chapter VIII above).

The fundamental group has two ends if and only if $\tilde{P} \simeq S^2$, and then P is homotopy equivalent to one of the four $S^2 \times E^1$-manifolds $S^2 \times S^1$, $S^2 \tilde{\times} S^1$, $RP^2 \times S^1$ or $RP^3 \sharp RP^3$.

Theorem 3. [Hi93] *Let P be a PD_3-complex such that $\pi = \pi_1(P)$ has a nontrivial finite normal subgroup N. Then either P is homotopy equivalent to $RP^2 \times S^1$ or π is finite.*

If $\pi_1(P)$ has a finitely generated infinite normal subgroup of infinite index then it has one end, and so P is aspherical. We shall discuss this case in the next section.

3. PD_3-groups

As an indecomposable, torsion free group is either infinite cyclic or has one end it follows from Tura'ev's Theorem that a PD_3-complex with torsion free fundamental group is the connected sum of aspherical PD_3-complexes with a 3-manifold with free fundamental group. Thus the study of such PD_3-complexes reduces largely to the study of PD_3-groups.

We can characterize the fundamental groups of 3-manifolds admitting one of the five geometries of aspherical Seifert type or which are finitely covered by surface bundles among all PD_3-groups in simple group-theoretic terms.

(As observed in Chapter II, there is a discrepancy between our definition of PD_n-complex and the definition of PD_n-group given in [Bi], which does not require that the group be finitely presentable. In this section we shall allow the broader notion).

Theorem 4. [Hi87] *Let G be a PD_3-group with a nontrivial almost finitely presentable normal subgroup N of infinite index. Then either*
(i) $N \cong Z$, G/N has one end and $H^2(G/N; Z[G/N]) \cong Z$; or
(ii) N is a surface group and G/N has two ends.
Proof. This follows from an LHSSS calculation. //

If N is a surface group G is the fundamental group of a 3-manifold which is double covered by the mapping torus of a surface homeomorphism.

With more effort, it can be shown that an almost finitely presentable subnormal subgroup is either infinite cyclic and normal or is a surface group and is normal in a subgroup of finite index [BH91].

It follows easily from Theorems 3 and 4 that a PD_3-complex P fibres over S^1 if and only if there is an epimorphism from $\pi = \pi_1(P)$ to Z whose kernel is almost finitely presentable, while it fibres over a closed surface if and only if either $\beta_1(E) > 0$ and π has an infinite cyclic normal subgroup with quotient torsion free or of order 2 or if π is cyclic. Moreover if π is infinite and is a nontrivial direct product then P is homotopy equivalent to the product of S^1 with a closed surface.

Theorem 5. [Hi85] *Let G be a PD_3-group with $\sqrt{G} \neq 1$. Then either*
(i) $h(\sqrt{G}) = 3$ and G is the group of an E^3- or Nil^3-manifold; or
(ii) $h(\sqrt{G}) = 2$ and G is the group of a Sol^3-manifold; or
(iii) $\sqrt{G}$ is a rank 1 abelian group.
Proof. Certainly $h(\sqrt{G}) \leq c.d.\sqrt{G} \leq 3$. Moreover $c.d.\sqrt{G} = 3$ if and only if $[G : \sqrt{G}]$ is finite [St77]. Hence G is virtually nilpotent if and only if $h(\sqrt{G}) = 3$. If $h(\sqrt{G}) = 2$ then $\sqrt{G}$ is locally abelian, and hence abelian. Moreover $\sqrt{G}$ must be finitely generated, for otherwise $c.d\sqrt{G} = 3$. Thus $\sqrt{G} \cong Z^2$ and case (*ii*) follows from Theorem 4. //

An equivalent formulation of this theorem is that either G is virtually poly-Z or the maximal elementary normal subgroup of G is a rank 1 abelian group or is trivial. For the maximal elementary amenable normal subgroup of G contains $\sqrt{G}$, and by Theorem I.5 it must be virtually solvable.

With an additional hypothesis (which may be redundant) it can be shown that in the rank 1 case G is the fundamental group of an $H^2 \times E^1$- or $\tilde{SL}$-manifold. The heart of the argument lies in showing that $G/\sqrt{G}$ has a subgroup of finite index and of finite cohomological dimension. A simpler argument might follow from a better understanding of the condition "$H^2(G/\sqrt{G}; Z[G/\sqrt{G}]) \cong Z$" provided by Theorem 4. As the argument as published in [Hi85] and [H] has a gap (in the claim that G_1/A is an HNN extension) we shall provide a correct version here.

Theorem 6. *Let G be a PD_3-group with $\sqrt{G}$ a rank 1 abelian group, and suppose that G has a subgroup of finite index with infinite abelianization. Then $\sqrt{G} \cong Z$ and $G/\sqrt{G}$ is virtually a PD_2-group.* //

Proof. Let $C = C_G(\sqrt{G})$. Then G/C is isomorphic to a subgroup of $Aut(\sqrt{G})$. Since $\sqrt{G}$ is isomorphic to a subgroup of Q the latter group is isomorphic to a subgroup of $Q^\times \cong Z^\infty \oplus (Z/2Z)$ and so G/C is abelian. If G/C is infinite then $c.d.C \leq 2$ [St77] and $\sqrt{G}$ is not finitely generated, so C is abelian, by Theorem 8.8 of [Bi], and hence G is solvable. But then $h(\sqrt{G}) > 1$, by the argument of Theorem VI.3, which is contrary to our hypothesis. Therefore G/C is isomorphic to a finite subgroup of $Q^\times$ and so has order at most 2. On passing to a subgroup of finite index if necessary we may assume that G/G' is infinite and that $C = G$, i.e., that $\sqrt{G} = \zeta G$.

Fix an epimorphism $\theta : G \to Z$ and let t be an element of $\theta^{-1}(1)$. If $\theta(\sqrt{G}) = nZ$ for some nonzero n then $\sqrt{G} \cong Z$ and $\theta^{-1}(nZ)$ is a subgroup of finite index in G which splits as a direct product $K \times Z$, where $K = \ker\theta$.

If $\theta(\sqrt{G}) = 0$ then $\sqrt{G} = \zeta K$. Since $c.d.K \leq 2$, by [St77], and G is not virtually poly-Z, K' is a nonabelian free group, by Theorem 8.8 of [Bi]. Hence $\sqrt{G} \cap K' = 1$. Since G is finitely generated $M = K/K'$ is finitely generated as a module over $Z[G/K] = Z[t, t^{-1}]$, and $\sqrt{G}$ maps injectively to a submodule which is annihilated by $t - 1$. Let $(t-1)^r$ be the highest power of $t-1$ that divides the order ideal of the $Z[t,t^{-1}]$-torsion submodule of M. Then $\sqrt{G}$ maps injectively to $N = M/(t-1)^r M$, which is finitely generated as an abelian group. In particular, we again find that $\sqrt{G} \cong Z$. There is an integer $m \geq 1$ such that $\sqrt{G}$ maps isomorphically onto an abelian group direct summand of mN. Let H be the preimage of mN in G. Then H has finite index in K and is normal in G, since mN is a sub-$Z[t,t^{-1}]$-module of N.

Now $H \cong (H/\sqrt{G}) \times \sqrt{G}$ and so $H/\sqrt{G}$ is isomorphic to a subgroup of H and has finite cohomological dimension. Since H is normal in G the subgroup G_1 generated by H and t is an extension of Z by H. Therefore $G_1/\sqrt{G}$ also has finite cohomological dimension. Since G_1 has finite index in G it is again a PD_3-group. Therefore $G_1/\sqrt{G}$ is a PD_2-group, by Theorem 9.11 of [Bi]. //

Corollary. *G is the fundamental group of a Seifert fibred 3-manifold.*

Proof. This follows from [Sc83] and Section 63 of [Z], as in [Hi85]. //

The geometric conclusions of Theorem 5 and 6 and the fact that 3-manifold groups are coherent suggest that Theorem 4 should remain true if the hypothesis that N be finitely presentable is weakened to "N is finitely generated".

There is at present no comparable characterization of the groups of H^3-manifolds, although it may be conjectured that these are exactly the PD_3-groups which have no noncyclic abelian subgroups.

PROBLEMS

1. Does $H^2(G;Z[G]) \cong Z$ imply that G is virtually a surface group? (This issue recurs throughout the book. For example, see Theorems II.11, V.13, VI.12, VI.13, VII.1 and X.4).

2. In particular, if M is a closed 4-manifold such that $\tilde{M} \simeq S^2$ is $\chi(M) \leq 0$?

3. If M is a compact complex analytic surface and $\pi_2(M) \cong Z$ must M be CP^2 or an elliptic surface or a ruled surface?

4. For which groups G is $H^2(G;Z[G]) = 0$? In particular, is this so if G is elementary amenable and $h(G) > 2$?

5. Can the hypothesis that $\pi_1(\hat{M})$ be FP_3 be removed from Theorem II.9.(*ii*)?

6. Clarify the criteria for a 4-manifold to be homotopy equivalent to the total space of an S^1-bundle over a PD_3-complex.

7. Extend the characterizations of 4-dimensional mapping tori and cartesian products.

8. Determine the homotopy types of the nonorientable closed 4-manifolds which are total spaces of surface bundles over RP^2 or are $S^2 \times E^2$- or $S^2 \times S^2$-manifolds.

9. Which extensions of surface groups by surface groups are realised by complex analytic surfaces?

10. Show that if G is the fundamental group of a closed aspherical 3-manifold then the group ring $Z[G]$ is coherent.

11. Let M be a closed 4-manifold with $\chi(M) = 0$ and such that $\pi = \pi_1(M)$ has an infinite amenable normal subgroup and $H^s(\pi;Z[\pi]) = 0$ for $s \leq 2$. Must M be aspherical?

12. Let M be an aspherical closed 4-manifold such that $\pi = \pi_1(M)$ has an

Typeset by $\mathcal{A}_{\mathcal{M}}\mathcal{S}$-TEX

amenable normal subgroup ν with $h.d_Q\nu \geq 3$. Must M be an infrasolvmanifold?

13. Determine the closed nonaspherical 4-manifolds M with torsion free elementary amenable fundamental group and $\chi(M) = 0$.

14. Let M be a closed 4-manifold with $\chi(M) = 0$ and such that $\pi = \pi_1(M)$ has a normal subgroup G of infinite index which is a PD_2-group. Must M be aspherical? If $\chi(G) < 0$ must π/G be virtually abelian?

15. Let M be a closed 4-manifold with $\chi(M) = 0$ and such that $\pi = \pi_1(M)$ has a torsion free abelian normal subgroup of rank 2 and also has a subgroup of finite index which has abelianization of rank at least 2. Must M be virtually simple homotopy equivalent to the total space of a torus bundle over a surface?

16. Is every S^2-bundle over an aspherical surface geometric?

17. Determine the fundamental groups of closed 4-manifolds with universal covering space $S^3 \times R$.

REFERENCES

[ABR92] Arapura, D., Bressler, P. and Ramachandran, M. On the fundamental group of a compact Kähler manifold,
Duke Math. J. 68 (1993), 477-488.

[AJ76] Auslander, L. and Johnson, F.E.A. On a conjecture of C.T.C.Wall,
J. London Math. Soc. 14 (1976), 331-332.

[Ba80] Barge, J. An algebraic proof of a theorem of J.Milnor,
in *Proceedings of the Topology Symposium, Siegen, 1979*,
Lecture Notes in Mathematics 788,
Springer-Verlag, Berlin - Heidelberg - New York (1980), 396-398.

[Ba80'] Barge, J. Dualité dans les revêtements galoisiens,
Invent. Math. 58 (1980), 101-106.

[BPV] Barth, W., Peters, C. and Van de Ven, A. *Compact Complex Surfaces*,
Ergebnisse der Mathematik und ihrer Grenzgebiete 3 Folge, Bd 4,
Springer-Verlag, Berlin - Heidelberg - New York - Tokyo (1984).

[Ba] Baues, H.J. *Obstruction Theory*,
Lecture Notes in Mathematics 628,
Springer-Verlag, Berlin - Heidelberg - New York (1977).

[Ba'] Baues, H.J. *Combinatorial Homotopy and 4-Dimensional Complexes*
De Gruyter Expositions in Mathematics 2,
Walter De Gruyter, Berlin - New York (1991).

[BB76] Baumslag, G. and Bieri, R. Constructable solvable groups,
Math. Z. 151 (1976), 249-257.

[Bi] Bieri, R. *Homological Dimension of Discrete Groups*,
Queen Mary College Mathematics Notes, London (1976).

[BH91] Bieri, R. and Hillman, J.A. Subnormal subgroups of 3-dimensional Poincaré duality groups, Math. Z. 206 (1991), 67-69.

[BS78] Bieri, R. and Strebel, R. Almost finitely presentable soluble groups,
Comment. Math. Helvetici 53 (1978), 258-278.

[BR84] Birman, J. and Rubinstein, J.H. One-sided Heegaard splittings and homeotopy groups of some 3-manifolds,
Proc. London Math. Soc. 49 (1984), 517-536.

[BO91] Boileau, M. and Otal, J.-P. Scindements de Heegaard et groupe des homéotopies des petites variétés de Seifert,
Invent. Math. 106 (1991), 85-107.

[Br72] Browder, W. Poincaré spaces, their normal fibrations and surgery,
Invent. Math. 17 (1972), 191-202.

Typeset by $\mathcal{A}\mathcal{M}\mathcal{S}$-TEX

[BG85] Brown, K.S. and Geoghegan, R. Cohomology with free coefficients of the fundamental group of a graph of groups, Comment. Math. Helvetici 60 (1985), 31-45.

[Ca89] Cairns, G. Compact 4-manifolds that admit totally umbilic metric foliations, preprint, Latrobe University (1989).

[Ca73] Cappell, S.E. Mayer-Vietoris sequences in Hermitean K-theory, in *Hermitean K-Theory and Geometric Applications* (edited by H.Bass), Lecture Notes in Mathematics 343, Springer-Verlag, Berlin - Heidelberg - New York (1973), 478-512.

[CG86] Cheeger, J. and Gromov, M. L_2-Cohomology and group cohomology, Topology 25 (1986), 189-215.

[CH90] Cochran, T.D. and Habegger, N. On the homotopy theory of simply-connected four-manifolds, Topology 29 (1990), 419-440.

[Co] Cohen, M.M. *A Course in Simple Homotopy Theory*, Graduate Texts in Mathematics 10, Springer-Verlag, Berlin - Heidelberg - New York (1973).

[CR77] Conner, P.E. and Raymond, F. Deforming homotopy equivalences to homeomorphisms in aspherical manifolds, Bull. Amer. Math. Soc. 83 (1977), 36-85.

[CS83] Culler, M. and Shalen, P.B. Varieties of group representations and splittings of three-manifolds, Ann. Math. 117 (1983), 109-146.

[Da83] Davis, M. Groups generated by reflections and aspherical manifolds not covered by Euclidean space, Ann. Math. 117 (1983), 293-325.

[DM85] Davis, J.F. and Milgram, R.J. A survey of the spherical space form problem, Math. Reports 2(1985), 223-283.

[DD] Dicks, W. and Dunwoody, M.J. *Groups acting on Graphs*, Cambridge studies in advanced mathematics 17, Cambridge University Press, Cambridge - New York - New Rochelle - Melbourne - Sydney (1989).

[EE69] Earle, C. and Eells, A fibre bundle description of Teichmüller theory, J. Diff. Geom. 3 (1969), 19-43.

[Ec92] Eckmann, B. Amenable groups and Euler characteristic, Comment. Math. Helvetici 67 (1992), 383-393.

[Ec93] Eckmann, B. Manifolds of even dimension with amenable fundamental group, preprint, ETH Zürich (February 1993).

[EL81] Eckmann, B. and Linnell, P.A. Poincaré duality groups of dimension two.II, Comment. Math. Helvetici 58 (1983), 111-114.

[EM80] Eckmann, B. and Müller, H. Poincaré duality groups of dimension two, Comment. Math. Helvetici 55 (1980), 510-520.

[EM82] Eckmann, B. and Müller, H. Plane motion groups and virtual Poincaré duality groups of dimension two, Invent. Math. 69 (1982), 293-310.

[Fa74] Farrell, F.T. The second cohomology group of G with coefficients $Z/2Z[G]$, Topology 13 (1974), 313-326.

[Fa75] Farrell, F.T. Poincaré duality and groups of type FP, Comment. Math. Helvetici 50 (1975), 187-195.

[FH81] Farrell, F.T. and Hsiang, W.C. The Whitehead group of poly-(finite or cyclic) groups,

J. London Math. Soc. 24 (1981), 308-324.

[FJ86] Farrell, F.T. and Jones, L.E. K-Theory and dynamics.I, Ann. Math. 124 (1986), 531-569.

[FJ88] Farrell, F.T. and Jones, L.P. The surgery L-groups of poly-(finite or cyclic) groups, Invent. Math. 91 (1988), 559-586.

[FJ89] Farrell, F.T. and Jones, L.E. A topological analogue of Mostow's rigidity theorem, J. Amer. Math. Soc. 2 (1989), 257-370.

[FJ90] Farrell, F.T. and Jones, L.E. Rigidity and other topological aspects of compact nonpositively curved manifolds, Bull. Amer. Math. Soc. 22 (1990), 59-64.

[FJ] Farrell, F.T. and Jones, L.E. *Classical Aspherical Manifolds*, CBMS Regional Conference Series 75, American Mathematical Society, Providence (1990).

[FJ93] Farrell, F.T. and Jones, L.E. Isomorphism conjectures in algebraic K-theory, J. Amer. Math. Soc. 6 (1993), 249-297.

[FJ93'] Farrell, F.T. and Jones, L.E. Topological rigidity for compact non-positively curved manifolds, in *Proceedings of Symposia in Pure Mathematics 54*, Part 3, American Mathematical Society (1993), 229-274.

[Fi] Filipkiewicz, R.O. *Four-Dimensional Geometries*, Ph.D thesis, University of Warwick (1984).

[FQ] Freedman, M.H. and Quinn, F. *Topology of 4-Manifolds*, Princeton University Press, Princeton (1990).

[Ga91] Gabai, D. Convergence groups are Fuchsian groups, Bull. Amer. Math. Soc. 25 (1991), 395-402.

[GM86] Geoghegan, R. and Mihalik, M.L. A note on the vanishing of $H^n(G;Z[G])$, J. Pure Appl. Alg. 39 (1986), 301-304.

[GK] Gordon, C. McA. and Kirby, R.C. (editors) *Four-Manifold Theory*, CONM 35, American Mathematical Society, Providence (1984).

[Go65] Gottlieb, D. A certain subgroup of the fundamental group, Amer. J. Math. 87 (1965), 840-856.

[Go68] Gottlieb, D. On fibre spaces and the evaluation map, Ann. Math. 87 (1968), 42-55.

[Go79] Gottlieb, D.H. Poincaré duality and fibrations, Proc. Amer. Math. Soc. 76 (1979), 148-150.

[Ha82] Habegger, N. Une variété de dimension 4 avec forme d'intersection paire et signature -8, Comment. Math. Helvetici 57 (1982), 22-24.

[HK88] Hambleton, I. and Kreck, M. On the classification of topological 4-manifolds with finite fundamental group, Math. Ann. 280 (1988), 85-104.

[HKT92] Hambleton, I., Kreck, M. and Teichner, P. Nonorientable four-manifolds with fundamental group of order 2, preprint, McMaster University (1992).

[HM86] Hambleton, I. and Madsen, I. Actions of finite groups on R^{n+k} with fixed point set R^k, Canadian Math. J. 38 (1986), 781-860.

[HM86'] Hambleton, I. and Madsen, I. Local surgery obstructions and space forms, Math. Z. 193 (1986), 191-214.

[HM78] Hambleton, I. and Milgram, R.J., Poincaré transversality for double covers, Canad. J. Math. 30 (1978), 1319-1330.

[HMTW] Hambleton, I., Milgram, R.J., Taylor, L.R. and Williams, B. Surgery with finite fundamental group, Proc. London Math. Soc. 56 (1988), 349-379.

[HW85] Hausmann, J.-C. and Weinberger, S. Caractérisques d'Euler et groupes fondamentaux des variétés de dimension 4, Comment. Math. Helvetici 60 (1985), 139-144.

[He77] Hendriks, H. Obstruction theory in 3-dimensional topology: an extension theorem, J. London Math. Soc. 16 (1977), 160-164. Corrigendum, *ibid.* 18 (1978), 192.

[He] Hendriks, H. *Applications de la théorie d'obstruction en dimension 3*, Bull. Soc. Math. France Memoire 53 (1977), 1-86.

[HL74] Hendriks, H. and Laudenbach, F. Scindement d'une équivalence d'homotopie en dimension 3, Ann. Sci. Ecole Norm. Sup. 7 (1974), 203-217.

[Hg40] Higman, G. The units of group rings, Proc. London Math. Soc. 46 (1940), 231-248.

[Hi85] Hillman, J.A. Seifert fibre spaces and Poincaré duality groups, Math. Z. 190 (1985), 365-369.

[H] Hillman, J.A. *2-Knots and their Groups*, Australian Mathematical Society Lecture Series 5, Cambridge University Press, Cambridge - New York - Melbourne (1989).

[Hi89] Hillman, J.A. A homotopy fibration theorem in dimension four, Topology Appl. 33 (1989), 151-161.

[Hi91] Hillman, J.A. Elementary amenable groups and 4-manifolds with Euler characteristic 0, J. Austral. Math. Soc. 50 (1991), 160-170.

[Hi91'] Hillman, J.A. On 4-manifolds homotopy equivalent to surface bundles over surfaces, Topology Appl. 40 (1991), 275-286.

[Hi91"] Hillman, J.A. On 4-manifolds homotopy equivalent to circle bundles over 3-manifolds Israel J. Math. 75 (1991), 277-287.

[Hi92] Hillman, J.A. Geometries on 4-manifolds, Euler characteristic and elementary amenable groups, in *Knots 90* (edited by A.Kawauchi), Walter de Gruyter Co., Berlin - New York (1992), 25-46.

[Hi93] Hillman, J.A. On 3-dimensional Poincaré duality complexes and 2-knot groups, Math. Proc. Cambridge Philos. Soc. 114 (1993) (to appear).

[Hi93'] Hillman, J.A. On 4-manifolds with finitely dominated covering spaces Proc. Amer. Math. Soc. (to appear).

[Hi94] Hillman, J.A. Minimal 4-manifolds for groups of cohomological dimension 2, Proc. Edinburgh Roy. Soc. (to appear).

[HL92] Hillman, J.A. and Linnell, P.A. Elementary amenable groups of finite Hirsch length are locally-finite by virtually-solvable, J. Austral. Math. Soc. 52 (1992), 237-241.

[HW89] Hillman, J.A. and Wilson, S.M.J. On the reflexivity of Cappell-Shaneson 2-knots, Bull. London Math. Soc. 21 (1989), 591-593.

[HR83] Hodgson, C. and Rubinstein, J.H. Involutions and isotopies of lens spaces, in *Knot Theory and Manifolds* (edited by D.Rolfsen), Lecture Notes in Mathematics 1144, Springer-Verlag, Berlin - Heidelberg - New York (1983), 60-96.

[In74] Inoue, M. On surfaces of class VII_0, Invent. Math. 24 (1974), 269-310.

[Iw49] Iwasawa, K. On some types of topological groups, Ann. Math. 50 (1949), 507-558.

[Jo79] Johnson, F.E.A. On the realisability of poly-surface groups, J. Pure Appl. Alg. 15 (1979), 235-241.

[K] Kaplansky, I.R. *Fields and Rings*, Chicago University Press, Chicago - London (1969).

[Ka75] Kato, M. Topology of Hopf surfaces, J. Math. Soc. Japan 27 (1975), 222-238. Erratum, *ibid.* 41 (1989), 173.

[KLR83] Kamashima, Y., Lee, K.B. and Raymond, F. The Seifert construction and its applications to infranilmanifolds, Quarterly J. Math. Oxford 34 (1983), 433-452.

[KKR92] Kim, M.H., Kojima, S. and Raymond, F. Homotopy invariants of nonorientable 4-manifolds, Trans. Amer. Math. Soc. 333 (1992), 71-83.

[KR90] Kim, M.H. and Raymond, F. The diffeotopy group of the twisted 2-sphere bundle over the circle, Trans. Amer. Math. Soc. 322 (1990), 159-168.

[KS] Kirby, R.C. and Siebenmann, L. *Foundational Essays on Topological Manifolds, Smoothings, and Triangulations*, Annals of Mathematics Study 88, Princeton University Press, Princeton (1977).

[Ko92] Kotschik, D. Remarks on geometric structures on compact complex surfaces, Topology 31 (1992), 317-321.

[Kr86] Kropholler, P.H. Cohomological dimensions of soluble groups, J. Pure Appl. Alg. 43 (1986), 281-287.

[KLM88] Kropholler, P.H., Linnell, P.A. and Moody, J.C. Applications of a new K-theoretic theorem to soluble group rings, Proc. Amer. Math. Soc. 104 (1988), 675-684.

[Kw86] Kwasik, S. On low dimensional s-cobordisms, Comment. Math. Helvetici 61 (1986), 415-428.

[KS88] Kwasik, S. and Schultz, R. Desuspensions of group actions and the ribbon theorem, Topology 27 (1988), 443-457.

[L] Laudenbach, F. *Topologie de la Dimension Trois: Homotopie et Isotopie*, Astérisque 12 (1974).

[Li91] Linnell, P.A. Zero divisors and group von Neumann algebras, Pacific J. Math. 149 (1991), 349-363.

[LYZ90] Li, J., Yau, S.-T. and Zheng, F. A simple proof of Bogomolov's theorem on Class VII_0 surfaces with $b_0 = 0$, Illinois J. Math. 34 (1990), 217-220.

[Mc] McCleary, J. *User's Guide to Spectral Sequences*, Mathematics Lecture Series 12,

Publish or Perish, Inc., Wilmington (1985).

[MTW76] Madsen, I., Thomas, C.B. and Wall, C.T.C. The topological spherical space form problem II. Existence of free actions, Topology 15 (1976), 375-382.

[Ma79] Matsumoto, T. On homotopy equivalences of $S^2 \times RP^2$ to itself, J. Math. Kyoto University 19 (1979), 1-17.

[MS86] Meeks, W.H.,III and Scott, G.P. Finite group actions on 3-manifolds, Invent. Math. 86 (1986), 287-346.

[Me84] Melvin, P. 2-Sphere bundles over compact surfaces, Proc. Amer. Math. Soc. 92 (1984), 567-572.

[Me88] Mess, G. The Seifert conjecture and groups which are coarse quasi-isometric to planes, preprint, UCLA (1988).

[Mi87] Mihalik, M.L. Solvable groups that are simply connected at ∞, Math. Z. 195 (1987), 79-87.

[Mi67] Milgram, R.J. The bar construction and abelian H-spaces, Ill. J. Math. 11 (1967), 241-250.

[Mi71] Milnor, J.W. *Introduction to Algebraic K-Theory*, Annals of Mathematics Study 72, Princeton University Press, Princeton (1971).

[Mi75] Milnor, J.W. On the 3-dimensional Brieskorn manifolds $M(p,q,r)$, in *Knots, Groups and 3-Manifolds* (edited by L.P.Neuwirth), Annals of Mathematics Study 84, Princeton University Press, Princeton (1975), 175-225.

[M] Montesinos, J.M. *Classical Tessellations and Three-Manifolds*, Springer-Verlag, Berlin - Heidelberg - New York (1987).

[Mo68] Mostow, G.D. Quasi-conformal mappings in n-space and the rigidity of hyperbolic space forms, Publ. Math. IHES 34 (1968), 53-104.

[Ne83] Neumann, W.D. Geometry of quasihomogeneous surface singularities, in *Singularities* (edited by P.Orlik), Proceedings of Symposia in Pure Mathematics 40, American Mathematical Society, Providence (1983), 245-258.

[N] Nicas, A.J. *Induction Theorems for Groups of Homotopy Manifold Structures*, Memoirs of the American Mathematical Society 267 (1982).

[NS85] Nicas, A. and Stark, C.W. K-Theory and surgery of codimension-two torus actions on aspherical manifolds, J. London Math. Soc. 31 (1985), 173-183.

[Ni42] Nielsen, J. Abbildungsklassen endlicher ordnung, Acta Math. 75 (1942), 23-115.

[Oh90] Ohba, K. On fiber bundles over S^1 having small Seifert manifolds as fibers, J. Fac. Sci. Tokyo 37 (1990), 659-702.

[Ol53] Olum, P. Mappings of manifolds and the notion of degree, Ann. Math. 58 (1953), 458-480.

[P] Pier, J. *Amenable Locally Compact Groups*, John Wiley, New York (1984).

[Pl80] Plotnick, S.P. Vanishing of Whitehead groups for Seifert manifolds with infinite fundamental group,

Comment. Math. Helvetici 55 (1980), 654-667.
[Pl82] Plotnick, S.P. Homotopy equivalences and free modules, Topology 21 (1982), 91-99.
[Pl84] Plotnick, S.P. Fibered knots in S^4 - twisting, spinning, rolling, surgery, and branching, in [GK], 437-459.
[PS87] Plotnick, S.P. and Suciu, A.I. Fibered knots and spherical space forms, J. London Math. Soc. 35 (1987), 514-526.
[Rg] Raghunathan, M.S. *Discrete subgroups of Lie Groups*, Ergebnisse der Mathematik 68, Springer-Verlag, Berlin - Heidelberg - New York (1972).
[Rn86] Ranicki, A. Algebraic and geometric splittings of the K- and L-groups of polynomial extensions, in *Transformation Groups, Poznan 1985* (edited by S.Jackowski and K.Pawalowski), Lecture Notes in Mathematics 1217, Springer-Verlag, Berlin - Heidelberg - New York (1986).
[Rn] Ranicki, A. *Algebraic L-Theory and Topological Manifolds*, Cambridge University Press, Cambridge - New York - Melbourne (1992).
[Ro] Robinson, D.J.S. *A Course in the Theory of Groups*, Graduate Texts in Mathematics 80, Springer-Verlag, Berlin - Heidelberg - New York (1982).
[Ro75] Robinson, D.J.S. On the cohomology of soluble groups of finite rank, J. Pure Appl. Alg. 6 (1975), 155-164.
[Ro68] Roos, J.E. Sur l'anneau maximal des fractions de AW^* algebras et des anneaux de Baer, C.R. Acad. Sci. Paris 266 (1968), 120.
[Ro84] Rosset, S. A vanishing theorem for Euler characteristics, Math. Z. 185 (1984), 211-215.
[Ru84] Rubermann, D. Invariant knots of free involutions of S^4, Top. Appl. 18 (1984), 217-224.
[Rb79] Rubinstein, J.H. On 3-manifolds which have finite fundamental group and contain Klein bottles, Trans. Amer. Math. Soc. 251 (1979), 125-137.
[Sc73] Scott, G.P. Finitely generated 3-manifold groups are finitely presentable, J. London Math. Soc. 6 (1973), 437-440.
[Sc83] Scott, G.P. There are no fake Seifert fibre spaces with infinite π_1, Ann. Math. 117 (1983), 35-70.
[Sc83'] Scott, G.P. The geometries of 3-manifolds, Bull. London Math. Soc. 15 (1983), 401-487.
[Sc85] Scott, G.P. Homotopy implies isotopy for some Seifert fibre spaces, Topology 24 (1985), 341-351.
[Se71] Serre, J.-P. Cohomologie des groupes discrets, in *Prospects in Mathematics*, Annals of Mathematics Study 70, Princeton University Press, Princeton (1971).
[Si71] Siebenmann, L.C. Topological manifolds, in *Proceedings of the International Congress of Mathematicians, Nice, 1970* vol. 2, Gauthier-Villars, Paris (1971), 133-163. *See* [KS].

[Si67] Siegel, J. Higher order cohomology operations in local coefficient theory, American J. Math. 89 (1967), 909-931.

[S] Spanier, E.H. *Algebraic Topology*, McGraw-Hill, New York (1966).

[Sp49] Specker, E. Die erste Cohomologiegruppe von Überlagerungen und Homotopie Eigenschaften dreidimensionaler Mannigfaltig keiten, Comment. Math. Helvetici 23 (1949), 303-333.

[St84] Stark, C.W. Structure sets vanish for certain bundles over Seifert manifolds, Trans. Amer. Math. Soc. 285 (1984), 603-615.

[St87] Stark, C.W. L-Theory and graphs of free abelian groups, J. Pure Appl. Alg. 47 (1987), 299-309.

[St92] Stark, C.W. Topological spaceform questions of mixed type, preprint, University of Florida, Gainesville (1992).

[St77] Strebel, R. A remark on subgroups of infinite index in Poincaré duality groups, Comment. Math. Helvetici 52 (1977), 317-324.

[Sw60] Swan, R.G. Periodic resolutions for finite groups, Ann. Math. 72 (1960), 267-291.

[Sw73] Swarup, G.A. On embedded spheres in 3-manifolds, Math. Ann. 203 (1973), 89-102.

[Th77] Thomas, C.B. On Poincaré 3-complexes with binary polyhedral fundamental group, Math. Ann. 226 (1977), 207-221.

[Th80] Thomas, C.B. Homotopy classification of free actions by finite groups on S^3, Proc. London Math. Soc. 40 (1980), 284-297.

[Tu90] Turaev, V.G. Three-dimensional Poincaré complexes: homotopy classification and splitting, Math. USSR Sbornik 67 (1990), 261-282.

[Ue90] Ue, M. Geometric 4-manifolds in the sense of Thurston and Seifert 4-manifolds I, J. Math. Soc. Japan 42 (1990), 511-540.

[Ue91] Ue, M. Geometric 4-manifolds in the sense of Thurston and Seifert 4-manifolds II, J. Math. Soc. Japan 43 (1991), 149-183.

[Wd67] Waldhausen, F. Eine Klasse von 3-dimensionaler Mannigfaltigkeiten I, Invent. Math. 3 (1967), 308-333; II, *ibid* 4 (1967), 87-117.

[Wd68] Waldhausen, F. On irreducible 3-manifolds which are sufficiently large, Ann. Math. 87 (1968), 56-88.

[Wd78] Waldhausen, F. Algebraic K-theory of generalized free products, Ann. Math. 108 (1978), 135-256.

[Wl65] Wall, C.T.C. Finiteness conditions for CW-complexes I, Ann. Math. 81 (1965), 56-69.

[Wl67] Wall, C.T.C. Poincaré complexes: I, Ann. Math. 86 (1967), 213-245.

[W] Wall, C.T.C. *Surgery on Compact Manifolds*, Academic Press, New York - London (1970).

[Wl76] Wall, C.T.C. Classification of hermitian forms.VI Group rings, Ann. Math. 103 (1976), 1-80.

[Wl78] Wall, C.T.C. Free actions of finite groups on spheres, in *Algebraic and Geometric Topology, Proceedings of Symposia in Pure Mathematics XXXII*, (edited by R.J.Milgram), American Mathematical Society, Providence (1978), Part 1, 115-124.

[Wl85] Wall, C.T.C. Geometries and geometric structures in real dimension 4 and complex dimension 2, in *Geometry and Topology* (edited by J.Alexander and J.Harer), Lecture Notes in Mathematics 1167, Springer-Verlag, Berlin - Heidelberg - New York (1985), 268-292.

[Wl86] Wall, C.T.C. Geometric structures on complex analytic surfaces, Topology 25 (1986), 119-153.

[We87] Weinberger, S. On fibering four- and five-manifolds, Israel J. Math. 59 (1987), 1-7.

[Wo] Wolf, J.A. *Spaces of Constant Curvature*, fifth edition, Publish or Perish Inc., Wilmington (1984).

[Yo92] Yoshikawa, K. Certain abelian subgroups of two-knot groups, in *Knots 90* (edited by A.Kawauchi), W. de Gruyter, Berlin - New York (1992), 231-240.

[Z] Zieschang, H. *Finite Groups of Mapping Classes of Surfaces*, Lecture Notes in Mathematics 875, Springer-Verlag, Berlin - Heidelberg - New York (1981).

[ZVC] Zieschang, H., Vogt, E. and Coldewey, H.D. *Surfaces and Planar Discontinuous Groups*, Lecture Notes in Mathematics 835, Springer-Verlag, Berlin - Heidelberg - New York (1980).

[Zn80] Zimmermann, B. Uber Gruppen von Homoömorphismen Seifertscher Faserräume und flacher Mannigfaltigkeiten, Manus. Math. 30 (1980), 361-373.

INDEX

Typeset by $\mathcal{A}\mathcal{M}\mathcal{S}$-TEX

Printed in the United States
By Bookmasters